SpringerBriefs in Molecular Science

Chemistry of Foods

Series Editor

Salvatore Parisi, Al-Balqa Applied University, Al-Salt, Jordan

The series Springer Briefs in Molecular Science: Chemistry of Foods presents compact topical volumes in the area of food chemistry. The series has a clear focus on the chemistry and chemical aspects of foods, topics such as the physics or biology of foods are not part of its scope. The Briefs volumes in the series aim at presenting chemical background information or an introduction and clear-cut overview on the chemistry related to specific topics in this area. Typical topics thus include:

- Compound classes in foods—their chemistry and properties with respect to the foods (e.g. sugars, proteins, fats, minerals, …)
- Contaminants and additives in foods—their chemistry and chemical transformations
- Chemical analysis and monitoring of foods
- Chemical transformations in foods, evolution and alterations of chemicals in foods, interactions between food and its packaging materials, chemical aspects of the food production processes
- Chemistry and the food industry—from safety protocols to modern food production

The treated subjects will particularly appeal to professionals and researchers concerned with food chemistry. Many volume topics address professionals and current problems in the food industry, but will also be interesting for readers generally concerned with the chemistry of foods. With the unique format and character of SpringerBriefs (50 to 125 pages), the volumes are compact and easily digestible. Briefs allow authors to present their ideas and readers to absorb them with minimal time investment. Briefs will be published as part of Springer's eBook collection, with millions of users worldwide. In addition, Briefs will be available for individual print and electronic purchase. Briefs are characterized by fast, global electronic dissemination, standard publishing contracts, easy-to-use manuscript preparation and formatting guidelines, and expedited production schedules.

Both solicited and unsolicited manuscripts focusing on food chemistry are considered for publication in this series. Submitted manuscripts will be reviewed and decided by the series editor, Prof. Dr. Salvatore Parisi.

To submit a proposal or request further information, please contact Dr. Sofia Costa, Publishing Editor, via sofia.costa@springer.com or Prof. Dr. Salvatore Parisi, Book Series Editor, via drparisi@inwind.it or drsalparisi5@gmail.com

More information about this subseries at http://www.springer.com/series/11853

Michele Barone · Rita Tulumello

Lathyrus sativus and Nutrition

Traditional Food Products, Chemistry and Safety Issues

 Springer

Michele Barone
Associazione 'Componiamo il Futuro'
(CO.I.F.)
Palermo, Italy

Rita Tulumello
Consulting and Global Service S.a.s.
Serradifalco, Caltanissetta, Italy

ISSN 2191-5407   ISSN 2191-5415   (electronic)
SpringerBriefs in Molecular Science
ISSN 2199-689X   ISSN 2199-7209   (electronic)
Chemistry of Foods
ISBN 978-3-030-59090-1   ISBN 978-3-030-59091-8   (eBook)
https://doi.org/10.1007/978-3-030-59091-8

This Springer imprint is published by the registered company Springer Nature Switzerland AG
The registered company address is: Gewerbestrasse 11, 6330 Cham, Switzerland

Contents

About the Authors

Michele Barone is an experienced consultant working in the field of food science and technology, and also in restoration chemistry. His work in food science focuses mainly on food packaging and correlated failures, and selected food products with a dedicated tradition (for instance, the Mediterranean Diet). More recently, he has written about food traceability systems in the field of European cheese products. Michele currently works at the Association 'Componiamo il Futuro' (CO.I.F.) in Palermo, Italy (sector: professional training). He has recently published *Dietary Patterns, Food Chemistry and Human Health, Quality Systems in the Food Industry*, and *Sicilian Street Foods and Chemistry. The Palermo Case Study* in the SpringerBriefs in Chemistry of Foods.

Rita Tulumello a biologist graduated in Industrial and Food Biotechnologies at the University of Palermo, Italy, works in the sector of chemical and microbiological analyses related to food and industrial (non-food) areas. Her main experiences concern the evaluation of microbiological and chemical profiles related to different food products, including cow milk cheeses, and environmental samples. In this ambit, Rita's first co-authored articles on food chemistry—*Studies on Determination of Antioxidant Activity of and Phenolic Content in Plant Products in India* (2000–2017), Sharma et al., and *Processed Meat and Polyphenols. Opportunities, Advantages and Difficulties*, Barbieri et al., *Journal of AOAC International* (2019)—concern the estimation of antioxidant properties ascribed to several polyphenols in vegetable tissues and their possible use in meat preparations.

Chapter 1
Lathyrus sativus: An Overview of Chemical, Biochemical, and Nutritional Features

Abstract One of the most important challenges in the modern world of food and beverage is the persistence of world hunger. Also, the surplus production of food in richer countries has determined important environmental problems. The problem should not be only the identification of the best food for human beings, but also the possibility of supplying this food in adequate amounts, with good nutritional intakes and in a sustainable way. In the ambit of crops, cereals are always the first choice. However, other vegetables can be considered with favour, including grain legumes. The use of domesticated legumes and pulses in human history is well documented worldwide. One of the most interesting grain legumes (because of its intrinsic resistance to drastic environmental conditions) is *Lathyrus sativus* (grass pea). Similarly to other grain legumes, it is historically reported to be used for the production of soups, unleavened breads, and other meals. On the other hand, grass pea may be questioned with concern to human and animal nutrition because of a peculiar neurotoxin associated with neurolathyrism disease. This chapter concerns chemical and biochemical properties of grass pea, with nutritional evaluations.

Keywords α-ODAP · β-ODAP · Grain legume · Grass pea · *Lathyrus sativus* · Neurolathyrism · Middle age

Abbreviations

ODAP	Diaminopropionic acid
FAO	Food and Agriculture Organization of the United Nations
DABA	*L*-2,4-diaminobutyric acid
β-ODAP	β-*N*-oxalyl-L-α,β-diaminopropionic acid
BAPN	β-aminopropionitrile toxin

© The Author(s), under exclusive license to Springer
Nature Switzerland AG 2020
M. Barone and R. Tulumello, *Lathyrus sativus and Nutrition*,
Chemistry of Foods, https://doi.org/10.1007/978-3-030-59091-8_1

1.1 An Introduction to Legumes: Modern and Historical Reasons

One of the most important challenges in the modern world of food and beverage is the persistence of world hunger. It has been reported that approximately 800 million people suffer hunger at present (Broughton et al. 2003): also, the surplus production of food in richer countries has determined important environmental problems, with real conflicts concerning water availability (Allan 1997; Allouche 2011; Falkenmark 2013; Shiva 2016). However, the basic key factors of this discussion can be summarised as follows (Norton et al. 2009):

(a) The current food surplus is abundant. Should the completed food supply worldwide be divided among the total world's population, it could be estimated that each person receives more than the minimum amount of needed nutrients each day. In other words, starvation is not a problem although the world's population has doubled constantly over the past four decades at least. In fact, food production is also increased in the same temporal period and probably faster than expected.

(b) The emergency is in terms of environmental risks on the one side (excessive soil exploitation) and of food hunger globally on the other side. In other terms, the faster the growth of agricultural production, the higher the risk of environmental damages worldwide. In addition, food hunger in this situation does not mean that food is not sufficient. The problem is the geographical and social difference between populations. After all, hunger is synonym of poverty (Barone and Pellerito 2020; Sharma et al. 2019). Moreover, many people suffer from chronical hunger in certain areas, while other populations consider food deficiency as a temporary problem. Interestingly, the localisation of most chronic hunger seems to be centred into rural areas worldwide (or suburbia in highly urbanised regions).

(c) From a rough geographical viewpoint, food hunger appears a feature observed in Asia, Africa (in particular when speaking of the Sub-Saharan area), and some regions of Latin America. Actually, this evaluation is not correct and the urbanisation has increased the diffusion of social inequalities in many world areas, including—necessarily—hunger.

(d) In addition, it should be observed that the main or unique activity in many world regions is agriculture. Consequently, even the free-of-charge donation of remarkable food amounts in selected areas with chronic food deficiency is not a good solution. On the contrary, the destruction of livelihoods and ancient traditions would be a medium-term consequence of this behaviour, on condition that it is economically sustainable.

It has to be also considered that food hunger is not a new phenomenon in the human history. On the contrary, it is probably one of the most ancient and recurring crises associated with anthropic life and activities (Augenti et al. 2006). As a result, the phenomenon has to be evaluated, studied, and re-evaluated again with a solid historical basis. After all, hunger and poverty are constant features in human civilisations, and the so-called street-food culture is not only the defence of cultural heritages and traditions in many regions, but also the demonstration that human problems may generate solutions such as 'poor' foods with good nutrient intakes (Barone and Pellerito 2020). It is known that food traditions should be considered as a cultural heritage able to define human history in many countries (Alfiero et al. 2017; Delgado et al. 2017; Maranzano 2014; Wilkinson 2004). Also, cultural cuisine suggests a notable correlation between food consumption on the one hand and social aggregation on the other hand. The relationship with historical famines in the Middle Age and other periods is also recognised, including also the identification of certain recipes as 'poor' or 'street' foods (Bell and Loukaitou-Sideris 2014; Booth and Coveney 2007; Chong and Eun 1992; Cortese et al. 2016; de Suremain 2016; Heinzelmann 2014; Helou 2006; Kollnig 2020; Larcher and Camerer 2015; Manning 2009; Maranzano 2014; McHugh 2015; Morton 2014; Patel et al. 2014; Pilcher 2017a, b; Sheen 2010; Simopoulus and Bhat 2000; Steven 2018; Webb and Hyatt 1988; Wilkins and Hill 2009).

Another reflection could be done when speaking of food culture on the one side and preservation of living species for food purposed on the other side (Metrick and Weitzman 1998; Osemeobo 1993; Sawicka et al. 2011; Spash and Hanley 1995; Tscharntke et al. 2012). In other words, food culture and heritage are continually evaluated, studies also with the aim of preserving it, no doubts about this fact and also food species, plants, and vegetables above all, are important and their defence is needed. Apparently, this reflection could have no relations with food heritage. On the contrary, many food cultures and traditions in the human history are the answer of civilisations against famine, hunger, and other emergencies impacting on humans and animals also. Consequently, biodiversity, food insecurity, and defence of food and beverage productions are parallel binaries. In addition, the problem of 'food wars' could be seen as a confirmation of this fact: food surplus (or deficiency) can be used as a weapon in the modern world, and it should be managed very carefully (Brussaard et al. 2010; Chappell and LaValle 2011; Elia 2000; Molotoks et al. 2017; Peh 2011; Sunderland 2011; Wittman et al. 2017).

So, the problem is to face modern emergency in terms of food hunger worldwide without damages to socio-economical structures of different populations. This discussion can be carried out in two ways at least: firstly, on the nutritional level (possibly, considering the current state of the art), and secondly with a retrospective evaluation of human history. After all, hunger is old as mankind (De Castro 1958; Fogel 2004; Mayer 1976; Vernon 2007).

1.2 The Nutritional Intake: What Is the Best Food?

With concern to the best choices in terms of foods and beverages, the difference between macronutrients and micronutrients should be always considered (Kennedy et al. 2003). The need of carbohydrates, fats, and protein on the one time should justify a plethora of different choices and food habits (or diets). Many choices, from Mediterranean Diet to low-fat or low-carbohydrate proposals or vegan/vegetarian lifestyles, could be discussed (Barone and Pellerito 2020). The possible use of vegan and similar foods could introduce a new important scenario in this ambit: the exclusive or main consumption of vegetables and fruits.

Actually, the preference for fruits and vegetables is not a new discovery in the human history. After all, many famines have been faced with the use of storable cereals and other vegetables. On the other side, famines are historically related to one or more high-energy nutrient vegetable foods and also historical outbreaks are associated with contaminated cereals such as ergotised wheat (Belser-Ehrlich et al. 2013; Biggs 1991; Malysheva et al. 2014; Urga et al. 2002). Substantially, the role of vegetables—generally in the form of stored crops but also as ingredients for prepared and cooked dishes—and fruits is one of the pillars of human history.

On the other side, the consumption of animal foods and ingredients for different prepared products has been the sign of social distinction and discrimination between rich and poor people for centuries (Gerbens-Leenes et al. 2010; Jiménez-Colmenero 2007; Popkin et al. 2001; Thang and Popkin 2004; Turrell and Kavanagh 2006). Only in the last decades, the consumption of red or white meat foods has been remarkable with interesting effects on human health in terms of increased cardiovascular diseases and augmented cancer risks (Campbell and Junshi 1994; Giovannucci et al. 1994; Trichopoulou and Lagiou 1997; Williamson et al. 2005). Anyway, animal foods and ingredients are not generally correlated with needed foods when speaking of hunger or famine periods. The economic importance of these foods and the associated use of natural resources[1] have to be taken into account seriously (Bosire et al. 2015; D'Silva and Webster 2017; Hoekstra 2012; Mekonnen and Hoekstra 2012; Sraïri et al. 2009; van Keulen 1975).

As a result, the problem should not be the identification of the best food for human beings only, but also the possibility of supplying this food in adequate amounts. This approach does not consider the nutritional importance of foods in terms of macro- and micronutrients: on the contrary, the point is to guarantee—where and when possible—the production of foods and beverages (Bhaskaran et al. 2006; De Haen and Réquillart 2014; Ebert 2014; van der Goot et al. 2016; Westgren 1999). In summary, food supply chains should be sustainable and possibly 'green' or 'eco-friendly' enough. In general, similar supply models are supposed and planned when speaking of vegetables and fruits, while the animal sector is complicated enough, although ethical considerations should suggest ethical choices as the first worrying in this ambit. The recent development of 'in vitro meat' foods or meat analogues in opposition to conventional meat foods has to be considered as a logical answer to this consumers' need. On the other hand, it should be considered that analogue meat preparations and their chemical composition are a declared attempt

---

[1]One single example: water for sanitisation purposes or for the realisation of cheeses is really notable if compared with water need for irrigation purposes.

to 'mime' the aspect of real meat, including colours. The use of soy leghemoglobin (a pigment able to transport oxygen turning itself into a red pigment, similarly to natural haemoglobin) has to be considered. Anyway, this and other innovations can notably increase the pressure on agricultural practices and yields, and we are not mentioning non-food applications such as ethanol for biodiesel applications (Aiking 2011; Bhat et al. 2015, 2017; D'Silva and Webster 2017).

In the ambit of crops, cereals are always the first choice (because they have an excellent importance when speaking of carbohydrates) (Barone and Pellerito 2020). However, food hunger is more and more challenging. Consequently, other vegetables can be considered with favour, including legumes, because of the good nutritional intake in terms of carbohydrates and also of protein molecules. With exclusive relation to nitrogen-based organic compounds, legumes can act as a good surrogate for animal proteins and this choice can be also studied in relation to historical traditions.

1.3 Legumes and Pulses in the Human History: The Beginnings and the Domestication

The use of legumes and pulses in human history is well documented worldwide (Aykroyd et al. 1982; Chang 1983; Deshpande 1992; Schuster-Gajzágó 2004; Stoddard et al. 2009). At present, the total production for this heterogeneous food group is reported to be about 77 million metric tons with a large diffusion in America, Africa, and Asian countries. However, this feature has been observed in European areas since the Middle Age at least, with an interesting development of legume and pulse productions after 1750. Interestingly, the use of forage legumes in Europe seems to show a certain decline after 1980 in favour or more intensive productions (Chorley 1981; Rochon et al. 2004). The use of lentils, lupines, peas, beans, and so on is related to the production of spreads, snacks, breakfast foods, meal ingredients, and soups. Probably, the historical success of pulses and legumes is related to notable protein contents (between 20 and 40%). For this reason, the use of legumes as protein addition in breads and pasta foods is an interesting option, while the possible use in gluten-free products (Barone and Pellerito 2020; Broomfield 2007; Fernández-Armesto 2002; Nene 2006; Pilcher 2017; Sathe 1996; Toussaint-Samat 2009; Sozer et al. 2017).

Anyway, legume crops have a broad enough origin from the geographical and historical viewpoint: the localisation comprehends Near Eastern and Central Asian regions and Mediterranean areas (Ljuština and Mikić 2008). One of the first sites with some evidence of cultivated grain legumes is reported to be Tell el-Kerkh (Syria), where traces seem to indicate ancient legume domestication. The most important findings in this area are related to the presence of legume seeds belonging to the following species:

(1) Bitter vetches
(2) Chickpeas, *garbanzos*, or Bengal grams (*Cicer arietinum*)
(3) Faba bean (*Vicia faba*)
(4) Grass peas (*Lathyrus sativus*)
(5) Lentils
(6) Peas.

Subsequently, the practice of cultivated grain legumes spread in the Old Europe from Asian areas through old fluvial routes such as Danube (intra-continental commerce) and the Mediterranean Sea. As a consequence, lentils, peas, and bitter vetches spread in the Neolithic Europe since 7500 B.C. at least (Santa Maria, Iberian Peninsula; Nevalı Çori, Turkey), while *Faba* legumes became the most used legume (or pulse?) after some time in some sites (Le Bras-Goude 2011; Ljuština and Mikić 2008; Mikić et al. 2011).

Interestingly, the diffusion of legumes in ancient European civilisations seems to be in connection with the use of fire in food preparation, as probably shown in many archaeological sites (Kaplan 2008). This observation may probably explain the transition from wild legumes to domesticated legumes worldwide, in strong connection with the concomitant cultivation and use of cereal grains (Kaplan 2008). This reflection is important enough because of digestion problems correlated to certain legumes and pulses by early hominids and the concomitant production of some neurotoxins and trypsin inhibitors, with a simple solution: denaturation by cooking. Substantially, it might be inferred that cereals and legumes have followed together the spreading of human 'fire cultures' at the same time, although the diffusion in many different areas of world continents may be difficult to be studied in detail. The example of *Phaseolus vulgaris* in pre-Columbian civilisations and archaeological sites (Guilá Naquits, Oaxaca, Mexico; Northeastern United States) demonstrates that maize (and cereals in general) was the first food component in these areas, while beans were the second food ingredient. Anyway, cereals and legumes were used together (Kaplan 2008). The use of other grain legumes such as chickpea (*Cicer arietinum*) in the old Europe of Middle Age is peculiar when speaking of certain food recipes which have 'survived' until the present age. The case of traditional Sicilian or Genoese products is made with (or containing) cereals or grain legumes (Abulafia 1978; Barbagli and Barzini 2010; Carrabino 2017; CNC 2014a, b, c, d, e; Dauverd 2006; Guigoni 2004; Pisano 2011). The use of *C. arietinum* in geographical areas where cereals are also well known and used is not a contradiction.

1.4 Legumes in the Old Europe: The Middle Age

The use of legumes in various European regions has been observed with peculiar interest in the Middle Age. In detail, the cereal/legume association (in relation to agricultural practices) has been demonstrated in certain areas such as Cap Vermell (Sant Julià de Lòria, Andorra) between the second and the twelfth centuries A.D.

(Peña-Chocarro and Peña 1999). This association with urbanised areas suggests a strong relationship between metallurgical or commercial activities on the one side and agricultural activities on the other side, meaning that ancient human communities were well balanced with concern to food production and non-food works. Other studies carried out especially in Italy and in France[2] (Bandini Mazzanti et al. 2005; Buonincontri et al. 2017; Cherubini 2000; Costantini et al. 1983; Esclassan et al. 2009; Montanari 2001) have demonstrated the coexistence of cereals and legumes in different ages, from the late Roman Empire to the late Middle Age. In other words, cereals and other plants such as *Pisum sativum, V. faba*, and *L. sativus* were often present and probably cultivated in good amounts. Interestingly, the production of neurotoxins or the persistency of digestion problems did not seem to be important enough for inhabitants: the demonstration is the presence of *L. sativus*, which is well known because of a particular toxin and the associated 'lathyrism' disease. This legume is no longer considered 'poor food' as in ancient ages. However, the use of this legume for peculiar recipes only and its persistence as a heritage of Middle Age traditions should be studied in detail. The same localisation of known recipes based on *L. sativus* in certain areas found in Morocco, Spain, France, and Italy (as an ideal line from the Maghreb area to Central Europe) can be seen as a recall of past memories. However, grass pea is not the only legume used for food preparations. Consequently, a premise should be made in relation to different pulses (or legumes?) and their classification before discussing grass pea.

1.5 Pulses or Legumes?

According to the Food and Agriculture Organization (FAO) of the United Nations, the current classification of pulses may be summarised as follows (The Global Pulse Confederation 2020; FAO 2015):

(1) Beans, dry
(2) Broad beans, dry
(3) Peas, dry
(4) Chickpeas
(5) Cowpeas, dry
(6) Pigeon peas

---

[2]Ferrara, 'The Mirror Pit' site, Italy; Vilarnau d'Amont, south-west of France; Miranduolo, Siena, Italy; Mura di Santo Stefano, Anguillara Sabazia, Rome, Italy; Tuscany region, Italy; and Classe, Ravenna, Italy.

(7)   Lentils
(8)   Bambara beans
(9)   Vetches
(10) Lupins
(11) Pulses nes
(12) Flour of pulses
(13) Bran of pulses.

According to the FAO, the name 'pulse' is correct only when speaking of crops for dry grain. This definition excludes other uses and green (fresh) legumes, or grains used solely for oil extraction or sowing treatments (The Global Pulse Confederation 2020; FAO 2015). As a result, legumes are a heterogeneous group of vegetable products, while 'pulses' are only a sub-typology of legumes. Consequently, the word 'legume' should be used broadly, while 'pulse' does not mean all possible legumes and this reflection is important enough.

In fact, the following words: 'legume', 'pulse', and also 'bean', are often used interchangeably, but meanings are really different (Anonymous 2020a):

(a)   A legume refers to any plant from the *Fabaceae* or *Leguminosae* family (more than 20,000 species). This simple definition includes all legume parts: stems, pods, and leaves.
(b)   The pulse corresponds to the edible see only, including lentils, peas, and beans. However, the FAO definition excludes some seeds from the 'pulses' group.
(c)   Consequently, 'beans' are only one of the possible legume seeds and pulses (where possible, according to the FAO definition for 'pulse').

The broad 'legume' definition concerns not only dry seeds (pulses) but also (Anonymous 2020a; FAO 2015; Legume Phylogeny Working Group et al. 2013):

(1)   Soybeans (*Glycine max* (L.) Merr.) and peanuts (*Arachis hypogaea*), mainly used for oil extraction
(2)   Clovers (*Trifolium* L.), leucaena (*Leucaena leucocephala* (Lam.) de Wit), and alfalfa (*Medicago sativa* L.), mainly cultivated for sowing forage
(3)   Ornamental species such as *Erythrina* L., *Lathyrus* L., *Lupinus* L., and *Acacia* Mill.
(4)   Medicinal plants such as *Glycyrrhiza* L.

By the viewpoint of classification specialists, the problem of legumes is that *Leguminosae* is usually subdivided into three groups, but the discussion is currently open: several researchers are accustomed to consider these families individually, without a common morphological origin (Legume Phylogeny Working Group et al. 2013). Consequently, the interest of food professionals for legumes is related to the fruit: dry or green seeds, without distinctions. The definition of other legume parts is not interesting in this ambit. As a result, the following discussions and chapters are related to 'grain legumes' or 'seeds' instead of pulses, legumes, or the general 'beans' name. In addition, this name 'grain legume' is good enough because of the suggested and historical relationship with 'grain cereals': both types have been domesticated and used for human and animal nutrition at the same time.

1.6 A Peculiar Grain Legume: Grass Pea

Figure 1.1 shows a simplified description concerning grain legumes, by the viewpoint of the FAO (the definition of 'pulses' should be mainly considered) on the left side and by the viewpoint of food technologists and consumers on the right side. In brief:

(1) The taxonomic description and classification of the most used and commonly known grain legumes start from legumes (*Fabaceae* or *Leguminosae*) with the subsequent discrimination into six sub-families: *Cercidoideae; Detarioideae; Duparquetioideae; Dialioideae; Caesalpinioideae*; and *Faboideae* (*Papilionoideae*), according to a new classification proposal (Legume Phylogeny Working Group et al. 2013; Sinou et al. 2020). One only of these sub-families comprehends grain legumes of our interest: *Faboideae*, containing 28 tribes. Five of these tribes are of our interest: *Phaseoleae* (containing also *Phaseolus vulgaris* L. or common beans); *Dalbergiae* (e.g. *Arachis hypogaea*); *Cicereae* (e.g. *Cicer arietinum*); *Trifolieae* (containing also *Medicago truncatula*), and the *Fabeae* (or *Vicieae*) tribe (Broich 2007; Macas et al. 2015; Minamioka and Imai 2009; Oskoueiyan et al. 2014; Schaefer et al. 2012). The last tribe contains five different genera (Angioi et al. 2009; Legume Phylogeny Working Group et al. 2013; Ochatt et al. 2016):

    (1.1) *Lathyrus* L. (generally named vetchlings)
    (1.2) *Lens* Mill. (commonly named lentils)
    (1.3) *Pisum* L. (commonly named peas)
    (1.4) *Vavilovia* Fed. (one species only: *Vavilovia formosa*)
    (1.5) *Vicia* L. (also named broad beans).

(2) The FAO classification of pulses can be useful enough when comparing it with the above-discussed presentation of grain legumes with some food interest. In general, it can be observed that the FAO recognises 13 groups:

(2.1) Beans, dry (the only mentioned genus is *Phaseolus* spp)

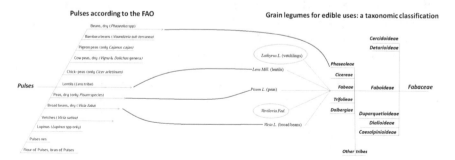

Fig. 1.1 A simplified description concerning grain legumes, by the viewpoint of the FAO (the definition of 'pulses' should be mainly considered) on the left side and by the viewpoint of food technologists and consumers on the right side

(2.2)   Broad beans, dry (*Vicia faba*)
(2.3)   Peas, dry (only *Pisum* spp.)
(2.4)   Chickpeas (only *Cicer arietinum*)
(2.5)   Cowpeas, dry (only *Vigna* and *Dolichos* genera)
(2.6)   Pigeon peas (only *Cajanus cajan*)
(2.7)   Lentils (*Lens* tribe)
 2.8)   Bambara beans (*Voandzeia subterranea*)
(2.9)   Vetches (*V. sativa*)
(2.10) Lupinus (*Lupinus* spp. only)
(2.11) Pulses nes (*Dolichos* spp.; *Canavalia* spp.; *Psophocarpus tetragonolobus*; *Cyamopsis tetragonoloba*; *Stizolobium* spp; *Pachyrhizus erosus*)
(2.12) Flour of pulses
(2.13) Bran of pulses.

Figure 1.1 shows that the correspondence between taxonomically correct classi-fications of legumes (of food interest) is not assured when speaking of 'pulses'. In detail, soybeans, peanuts, clovers, alfalfa, and other previously mentioned legumes are not cited. Moreover, some peculiar genus or species are not considered. This reflection concerns also:

(a)   *Vavilovia* Fed. genus (one species only: *V. Formosa*, a peculiar species under danger of extinction and with a few scientific studies only) and
(b)   *Lathyrus* genus, and more specifically a legume crop used for human nutrition for centuries: *Lathyrus sativus*, similar to certain *Vicia*, *Lens*, and *Pisum* species.

The use of *L. sativus* in human nutrition is well documented, and several traditions should be considered carefully. On the other side, this legume has a 'bad' reputation because of a particular neurotoxin associated with its excessive consumption. We would desire to give a good overview both of the historical and nutritional importance of this legume on the one side, and to discuss critically the health and safety worries associated with *L. sativus* is used as main or important ingredient of human foods.

L. sativus, commonly named grass pea or sweet pea, is only one of 160 *Lathyrus* species (Kumar et al. 2013), and it is also studied for both historical reasons and toxicological properties. This crop is really interesting because of its capability to survive in extremely difficult environments (where only fragile ecosystems could be observed). Difficult environmental conditions comprehend not only excessive heat, but also water stagnation, famine, and drought (Ben-Brahim et al. 2002; Campbell 1997; Cocks et al. 2000; Kumar et al. 2011a, b, 2013). It can be found in all continental areas without exceptions at present. In general, grass pea is found in the north and the Sub-Saharan area of Africa, South Asia countries, Southern Europe, and South America also. More recently, a limited presence has been evaluated in Australia also because of the need of pulses, in particular when speaking of fine-textured soils with pH ranging between extreme low and high values. In relation to Australia only, the introduction of different legumes such as lupines and subterranean clovers is reported for different soils with sandy features or fine texture. However, *L. sativus* (and also

L. cicera, known as chickling vetch) have been proposed in recent years (Cocks et al. 2000).

In this situation, it is reported that this annual legume crop—able to grow in the cool season—can have an economic and ecological importance worldwide. In relation to eating purposes only, grass pea is cited in Bangladesh, Ethiopia, India, Nepal, and Pakistan (Kumar et al. 2013). In addition, *L. sativus* is reported to be extensively present in European countries such as Southern Germany, Spain, Portugal, Crete, Southern Russia, Rhodes, and Cyprus. The Middle East and the Maghreb area are also important regions where grass pea can be found (Campbell 1997).

The relationship with human cultivations for nutrition has been postulated since the Neolithic Era in Europe, at least in the Balkan Peninsula, while the earliest traces have been found in Western Asia and the Indian Peninsula (Allchin 1969; Duke et al. 1981; Kislev 1989; Saraswat 1980). As a result, *Lathyrus* genus seems to be a constant presence in the human history similarly to other legumes.

The taxonomic description and biological features of grass pea are not among the objectives of this book. On the other hand, the economic and nutritional importance of *L. sativus* has to be discussed with an eye on important safety issues. After this premise, it should be also recognised that grass pea can be defined and named in many ways. The following and non-exhaustive list shows only several of possible named for this legume (Fig. 1.4), possibly in relation to the continental area (Campbell 1997; Lambein et al. 2010):

(1) Asia: *khesari* (Bangladesh); *san lee dow* (China); *batura, kesare, chattra matur,* chickling vetch, chickling pea, *karas, karil, lang,* etc. (India)
(2) Europe: *fovetta,* dog-toothed pea (Cyprus); *gesse blanche, lentille d'Espagne* (France); *saat-platterbse* (Germany); *chícharo* (Portugal); *almortas, guijas, muelas,* or *titos* (Spain); *cicerchia coltivata, pisello bretonne, pisello cicerchia* (Italy)
(3) Africa: *guaya, Sabberi* (Ethiopia); *gilban(eh)* (Sudan)
(4) South America: *frijol gallinazo, garbanuzo* (Venezuela).

As a result, the same legume may be differently found in different countries, and this circumstance may explain partially the confusion concerning *L. sativus* varieties, especially when speaking of food recipes containing this food.

In general, grass pea is a palatable food, and this feature is not generally the notable protein content (between 25 and 30%, with a remarkable presence of lysine and a reduced amount of sulphur amino acids). In addition, it is reported to contain more B vitamins and iron than other grain legumes. This circumstance and the good resistance to drastic environmental conditions such as drought and natural disasters have been repeatedly considered when speaking of possible introduction for grass pea in 'new' continental areas such as Australia. Moreover, fat contents are low enough, and less of 60% of fatty acids are reported to be unsaturated. Also, this food may be also used as a high-protein content animal feed. It should also be noted that certain ethical cultures do not allow the consumption of animal foods (and consequently animal protein) in certain year periods due to religion prohibitions (in Ethiopia at least). Consequently, grass pea can be a good protein surrogate.

On the other side, there are several disadvantages concerning *L. sativus*:

(a) First of all, the deficiency of vitamins A and C would need to be considered (Cocks et al. 2000; Spencer and Palmer 2003).

(b) The fibre content for this legume is reported to be lower than other species such as narrow-leafed lupines. This fact is not necessarily a problem, with the concomitant low amount of fat matter. On the other hand, modern nutrition recommends a good supplementation of fibres (Cocks et al. 2000).

(c) Finally, the consumption of grass pea is not a 'recommended' use because of the toxic properties of this legume. The associated disease is called 'neurolathyrism', even if this event may be also found together with another disease, the *konzo* illness ('tied legs), caused by ingestion of roots of *Manihot esculenta* (cassava) roots. This additional disease is the signal that cassava roots contain cyanogens (Gresta et al. 2014; Lambein et al. 2010; Nzwalo and Cliff 2011). In exclusive relation to grass pea only, the problem is the presence of a peculiar neurotoxin: β-*N*-oxalyl-L-α,β-diaminopropionic acid (β-ODAP). This toxin (Fig. 1.2) is present in all tissues of the legume, and it is able to cause the paralysis of lower limbs, also called neurolathyrism or lathyrism (Campbell and Junshi 1994). It has been reported that β-ODAP might be an important antinutritional factor, although further studies would be needed. An alternative isomer, the α-form, is considered less dangerous (Fig. 4.1).

(d) In addition, the profile of condensed tannins and total phenolics may suggest astringency with associated bitterness. Also, grass pea is reported to have some enzymatic inhibitory activities with peculiar reference to trypsin and chymotrypsin (Deshpande and Campbell 1992).

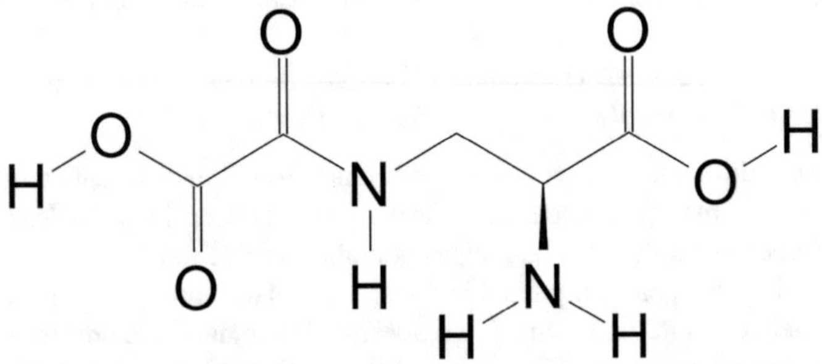

Fig. 1.2 Neurolathyrism disease is strongly associated with a peculiar neurotoxin: ß-*N*-oxalyl-L-α,β-diaminopropionic acid (β-ODAP). This toxin—also named β-N-oxalyl-ammo-L-alanine (BOAA) or dencichin, molecular weight: 176, molecular formula: $C_5H_8N_2O_5$—is present in all grass pea tissues, and it is able to cause the paralysis of lower limbs. It has been reported that β-ODAP might be an important antinutritional factor, although further studies would be needed. ODAP is also present as the α-isomer (Fig. 4.1) which is believed to be less toxic than β-ODAP (Fig. 4.1). Toxicity depends mainly on the β/α ratio (in nature, approximately 95:5, but this ratio can be modified in different conditions with different results)

It has been reported many times that the main and most worrying safety and health concern is the occurrence of neurolathyrism. Consequently, the associated neurotoxin should be discussed in detail.

1.7 Lathyrism and β-ODAP

The 'lathyrism' disease is not specifically related to *L. sativus* species only. On the contrary, the *Lathyrus* genus is associated with this phenomenon. It has to be considered that the 'lathyrism' disease implies generally paralysis, and it is not observed in humans only: animals, specifically ruminants and monogastric species, can be affected in the same way (Hanbury et al. 1999). Anyway, the responsible toxin—β-ODAP—was identified after 1960 (Murti et al. 1964).

In addition, lathyrism concerns two different paralysis diseases: osteolathyrism and neurolathyrism. The first phenomenon is associated with bones (skeletal deformities), and it is generally reported to be caused by the consumption of four *Lathyrus* species: *L. odoratus*, *L. hirsutus*, *L. pusillus*, and *L. roseus* (Hanbury et al. 1999). In this ambit, grass pea is not specifically correlated with osteolathyrism, although some reported studies imply that the presence of the β-aminopropionitrile toxin (BAPN, responsible for this disease) in *L. sativus* and *L. cicera* also could be the cause of emerging osteolathyrism symptoms in subjects already suffering of chronic neurolathyrism (Hanbury et al. 1999). BAPN (Fig. 1.3) is discussed here because neurolathyrism has been sometimes correlated with osteolathyrism, in spite of different symptoms (Fig. 1.4).

On the other side, neurolathyrism concerns more than a single disease only. In addition, it should be considered that the heavy consumption of grass pea is associated with this phenomenon, and also some *Vicia* and *Lathyrus* species have been

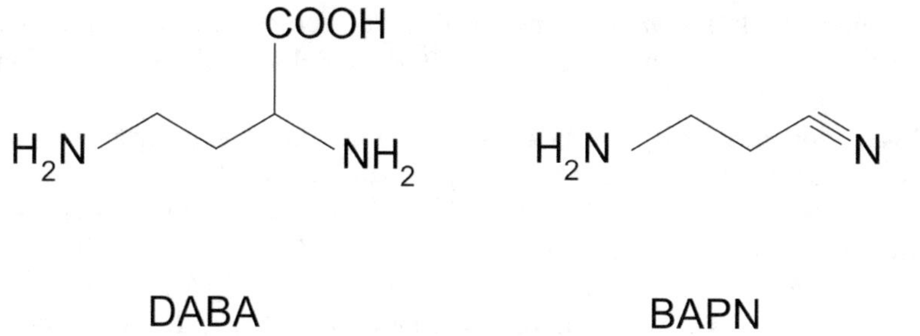

DABA   BAPN

Fig. 1.3 Lathyrism concerns two different paralysis diseases: osteolathyrism—sometimes associated with the presence of the β-aminopropionitrile toxin (BAPN) in *L. sativus* and *L. cicera*—and neurolathyrism. The last disease should be subdivided into two categories: the first one, mainly associated with *L. sylvestris* and *L. latifolius*, is caused by the toxic action of *L*-2,4-diaminobutyric acid (DABA). DABA is not detected in *L. sativus* and *L. cicera*. For this reason, the second neurolathyrism form and the related nature of the associated toxin, β-ODAP, have been extensively studied so far

Lathyrus sativus. One grain legume... Many different names

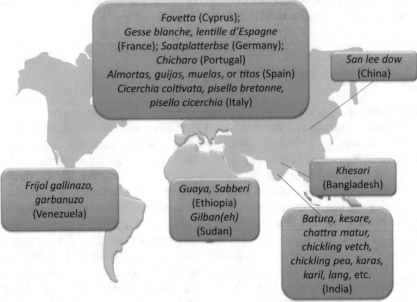

Fovetta (Cyprus);
Gesse blanche, lentille d'Espagne
(France); Saatplatterbse (Germany);
Chícharo (Portugal)
Almortas, guijas, muelas, or titos (Spain)
Cicerchia coltivata, pisello bretonne,
pisello cicerchia (Italy)

San lee dow
(China)

Frijol gallinazo,
garbanuzo
(Venezuela)

Guaya, Sabberi
(Ethiopia)
Gilban(eh)
(Sudan)

Khesari
(Bangladesh)

Batura, kesare,
chattra matur,
chickling vetch,
chickling pea, karas,
karil, lang, etc.
(India)

Fig. 1.4 It should be also recognised that grass pea can be defined and named in many ways. This picture shows several names for this legume, possibly in relation to the continental area (Campbell 1997; Lambein et al. 2010). The existence of different names can be challenging enough at first sight

involved. The general disease implies the perceived weakness of hind limbs and the concomitant paralysis of muscles. In particular, the human effects have been detailed as follows: initially, the subject feels painful spasms in lower limbs (muscles), with concomitant weakness and the subsequent spastic paraplegia in chronic form. The result can be often the complete and generally irreversible paralysis of legs (Hanbury et al. 1999).

In addition, neurolathyrism should be subdivided into two categories depending on the consumption of peculiar Lathyrus species at least. The first of these typologies, mainly associated with *L. sylvestris* and *L. latifolius* species only, is caused mainly by the toxic action of *L*-2,4-diaminobutyric acid (DABA, Fig. 1.3). However, the incidence and related clinic importance of DABA is relatively low if compared with neurolathyrism caused by consumption of grass pea, *L. cicera*, *L. clymenum*, and *L. ochrus* species. In fact, the most part of neurolathyrism episodes with exceptional importance from the safety viewpoint in humans and animals also depend on the presence of another toxin, while DABA is not detected in *L. sativus* and *L. cicera*. For this reason, the second neurolathyrism form is the most studied disease, and the related nature of the associated toxin, β-ODAP, has been extensively researched in recent years. In fact, the non-protein amino acid 3-(-*N*-oxalyl)-*L*-2,3-diamino propionic acid or β-N-oxalylamino-*L*-alanine (β-ODAP; Fig. 1.2) is detected in grass pea, *L. cicera, L. clymenum,* and *L. ochrus* species (Hanbury et al. 1999). Interestingly, the α-ODAP isomer is not associated with neurolathyrism (Gresta et al. 2014).

Recent studies have clarified the β-ODAP role in paralysis of muscles. Probably, β-ODAP is confused by the human or animal nervous system with a glutamate analogue, with resulting strong bonds between β-ODAP and α-amino-3-hydroxy-5-methyl-4-isoxazolepropionic acid-type glutamate receptors. In this way, neurons can suffer excitotoxic degeneration, and possibly other neurotoxic effects can be observed, probably with remarkable damages in young humans (Rao et al. 1964). On the other hand, β-ODAP is believed to have a protective role in plants against insects and yeasts, probably because humans and plants also have a distinctive lack of zinc ions (Lambein and Kuo 1997).

Anyway, β-ODAP is only the most known and dangerous of antinutritional factors correlated with grass pea. Other antinutritional compounds, as mentioned above, include tannins, trypsin, and chymotrypsin inhibitors, amylase inhibitors, phytic acid, oligosaccharides, etc. (Deshpande and Campbell 1992; Hanbury et al. 1999; Rao et al. 1964; Urga et al. 1995). The problem is that none of remaining molecules can be worrying as β-ODAP. It has to be also noted that this toxin is not found in *L. cicera* with high quantities if compared with grass pea. For this reason, the association between β-ODAP/neurolathyrism and *L. sativus* is the most known and studied phenomenon in this ambit. Consequently, safety and health worries concerning neurolathyrism imply tacitly grass pea and vice versa. The next chapters are dedicated to different aspects of grass pea, including also the lathyrism problem and possible solutions. For example, the association of *L. sativus* and cereals—with high amounts of sulphur amino acids, if compared with grass pea)—can 'adjust' the low quantity of cystine and methionine and lower β-ODAP toxicity at the same time (Gresta et al. 2014; Hillocks and Maruthi 2012). In addition, food processing—cooking, roasting, fermentation, etc.—can be interesting methods when speaking of β-ODAP reduction, in connection with soaking technique (β-ODAP is soluble in water, and its content may be lowered by means of simple aqueous 'washing' and consequent extraction) (Akalu et al. 1998). Consequently, grass pea can be re-evaluated in the current world of cereals and legumes with interesting results.

References

Abulafia D (1978) Pisan commercial colonies and consulates in twelfth-century Sicily. Eng Historical Rev 93(366):68–81

Aiking H (2011) Future protein supply. Trends Food Sci Technol 22(2–3):112–120. https://doi.org/10.1016/j.tifs.2010.04.005

Akalu G, Johansson G, Nair BM (1998) Effect of processing on the content of β-N-oxalyl-α, β-diaminopropionic acid (gb-ODAP) in grass pea (Lathyrus sativus) seeds and flour as determined by flow injection analysis. Food Chem 62(2):233–237. https://doi.org/10.1016/S0308-814 6(97)00137-4

Alfiero S, Giudice AL, Bonadonna A (2017) Street food and innovation: the food truck phenomenon. Brit Food J 119(11):2462–2476. https://doi.org/10.1108/BFJ-03-2017-0179

Allan JA (1997) 'Virtual water': a long term solution for water short Middle Eastern economies? School of Oriental and African Studies Water Issues Study Group, University of London, Occasional Papers. School of Oriental and African Studies, University of London, London, pp 24–29. Available https://www.soas.ac.uk/water/publications/papers/file38347.pdf. Accessed 08 July 2020

Allchin FR (1969) Early cultivated plants in India and Pakistan. In: Ucko J, Dimbleby GW (eds) The domestication and exploitation of plants and animals. Duckworth, London

Allouche J (2011) The sustainability and resilience of global water and food systems: Political analysis of the interplay between security, resource scarcity, political systems and global trade. Food Policy 36(1):S3–S8. https://doi.org/10.1016/j.foodpol.2010.11.013

Angioi SA, Desiderio F, Rau D, Bitocchi E, Attene G, Papa R (2009) Development and use of chloroplast microsatellites in *Phaseolus* spp. and other legumes. Plant Biol 11, 4:598–612. https://doi.org/10.1111/j.1438-8677.2008.00143.x

Anonymous (2020a) The nutrition source—legumes and pulses. Harvard T.H. Chan School of Public Health, Boston. Available https://www.hsph.harvard.edu/nutritionsource/legumes-pulses/. Accessed 10 July 2020

Augenti A, Bondi M, Carra M, Cirelli E, Malaguti C, Rizzi M (2006) Indagini archeologiche a Classe (scavi 2004): primi risultati sulle fasi di età altomedievale e dati archeobotanici. In: Proceedings of the fourth Congresso Nazionale di Archeologia Medievale. Scriptorium dell'abbazia, Abbazia di San Galgano, Chiusdino, Siena, Italy, 26–30 September 2006. All'Insegna del Giglio, Borgo San Lorenzo (FI), pp 153–161. ISBN 88-7814-469-X

Aykroyd WR, Doughty J, Walker AF (1982) Legumes in human nutrition, vol 20. FAO Food and Nutrition Paper No 20. Food and Agriculture Organization of the United Nations, Rome

Bandini Mazzanti M, Bosi G, Mercuri AM, Accorsi CA, Guarnieri C (2005) Plant use in a city in Northern Italy during the late Mediaeval and Renaissance periods: results of the archaeobotanical investigation of "The Mirror Pit" (14th–15th century a.d.) in Ferrara. Veget Hist Archaeobot 14:442–452. https://doi.org/10.1007/s00334-005-0082-y

Barbagli A, Barzini S (2010) Pane, pizze e focacce. Giunti Editore, Florence

Barone M, Pellerito A (2020) Sicilian Street Foods And Chemistry—the Palermo Case study. Springer International Publishing, Cham. https://doi.org/10.1007/978-3-030-55736-2

Bell JS, Loukaitou-Sideris A (2014) Sidewalk informality: an examination of street vending regulation in China. Int Plan Stud 19(3–4):221–243. https://doi.org/10.1080/13563475.2014.880333

Belser-Ehrlich S, Harper A, Hussey J, Hallock R (2013) Human and cattle ergotism since 1900: symptoms, outbreaks, and regulations. Toxicol Ind Health 29(4):307–316. https://doi.org/10.1177/0748233711432570

Ben-Brahim N, Salho A, Chtorou N, Combes D, Marrakcho M (2002) Isozymic polymorphism and phylogeny of 10 Lathyrus species. Genet Res Crop Evol 49:427–436. https://doi.org/10.1023/A:1020629829179

Bhaskaran S, Polonsky M, Cary J, Fernandez S (2006) Environmentally sustainable food production and marketing. Brit Food J 108(8):677–690. https://doi.org/10.1108/00070700610682355

Bhat ZF, Kumar S, Fayaz H (2015) In vitro meat production: challenges and benefits over conventional meat production. J Integr Agric 14(2):241–248. https://doi.org/10.1016/S2095-311 9(14)60887-X

Bhat ZF, Kumar S, Bhat HF (2017) In vitro meat: a future animal-free harvest. Crit Rev Food Sci Nutr 57(4):782–789. https://doi.org/10.1080/10408398.2014.924899

Biggs RD (1991) Ergotism and other mycotoxicoses in ancient Mesopotamia. Aula Orientalis 9(1–2):15–21

Booth SL, Coveney J (2007) Survival on the streets: prosocial and moral behaviors among food insecure homeless youth in Adelaide. South Australia. J Hunger Environ Nutr 2(1):41–53. https://doi.org/10.1080/19320240802080874

Bosire CK, Ogutu JO, Said MY, Krol MS, de Leeuw J, Hoekstra AY (2015) Trends and spatial variation in water and land footprints of meat and milk production systems in Kenya. Agric Ecosyst Environ 205:36–47. https://doi.org/10.1016/j.agee.2015.02.015

Broich SL (2007) New combinations in North American *Lathyrus* and *Vicia* (*Fabaceae: Faboideae: Fabeae*). Madroño 54(1):63–71. https://doi.org/10.3120/0024-9637(2007)54%5b63: ncinal%5d2.0.co;2

Broomfield A (2007) Food and cooking in Victorian England: a history. Praeger Publishers, Westport

Broughton WJ, Hernández G, Blair M, Beebe S, Gepts P, Vanderleyden J (2003) Beans are model food legumes. Plant Soil 225:55–128. https://doi.org/10.1023/A:1024146710611

Brussaard L, Caron P, Campbell B, Lipper L, Mainka S, Rabbinge R, Babin D, Pulleman M (2010) Reconciling biodiversity conservation and food security: scientific challenges for a new agriculture. Curr Opin Environ Sustain 2(1–2):34–42. https://doi.org/10.1016/j.cosust.2010.03.007

Buonincontri MP, Pecci A, Di Pasquale G, Ricci P, Lubritto C (2017) Multiproxy approach to the study of Medieval food habits in Tuscany (central Italy). Archaeol Anthropol Sci 9(4):653–671. https://doi.org/10.1007/s12520-016-0428-7

Campbell CG (1997) Grass pea, *Lathyrus sativus* L. Promoting the conservation and use of underutilized and neglected crops. Rome, Italy: Institute of Plant Genetics and Crop Plant Research, Rome, and Germany/International Plant Genetic Resources Institute, Gatersleben, Nr 18. p 92

Campbell TC, Junshi C (1994) Diet and chronic degenerative diseases: perspectives from China. Am J Clin Nutr 59(5):1153S–1161S. https://doi.org/10.1093/ajcn/59.5.1153S

Carrabino D (2017) 14 oratories of the Compagnie of Palermo: Sacred spaces of rivalry. In: Bullen Presciutti D (ed) Space, place, and motion: locating confraternities in the Late Medieval and Early Modern City. Koninklijke Brill Nv (Brill), Leiden, pp 344–371. https://doi.org/10.1163/978900 4339521_016

Chang TT (1983) The origins and early cultures of the cereal grains and food legumes. In: Keightley DN (ed) The origins of Chinese civilization. University of California Press, Berkeley, Los Angeles, and London, pp 65–94

Chappell MJ, LaValle LA (2011) Food security and biodiversity: can we have both? An agroecological analysis. Agric Hum Values 28(1):3–26. https://doi.org/10.1007/s10460-009-9251-4

Cherubini G (2000) L'approvvigionamento alimentare delle città toscane tra il XII e il XIV secolo. Rivista di storia dell'agricoltura 40:33–52

Chong HK, Eun VLN (1992) 4 backlanes as contested regions: construction and control of physical. In: Huat CB, Edwards N (eds) Public space: design, use and management. Singapore University Press, Singapore

Chorley GPH (1981) The agricultural revolution in northern Europe, 1750–1880: nitrogen, legumes, and crop productivity. Econ History Rev 34(1):71–93. https://doi.org/10.2307/2594840

CNC (2014a) Accordo tra Comune di Palermo e Consiglio Nazionale dei Chimici (CNC) per la partecipazione alla realizzazione di un sistema di salvaguardia e garanzia della tradizione gastronomica palermitana', Prot. 646/14/cnc/fta. Consiglio Nazionale dei Chimici (CNC), Rome. Available https://www.chimicifisici.it/wp-content/uploads/2018/10/20131210_accordo_firmato_dal_ Presidente_del_CNC.pdf. Accessed 7 April 2020

CNC (2014b) DOMANDA DI REGISTRAZIONE - Art. 8 - Regolamento (UE) n. 1151/2012 del Parlamento Europeo e del Consiglio del 21 novembre 2012 sui regimi di qualità dei prodotti agricoli e alimentari - "ARANCINA". Annex to the document 'Accordo tra Comune di Palermo e Consiglio Nazionale dei Chimici (CNC) per la partecipazione alla realizzazione di un sistema di salvaguardia e garanzia della tradizione gastronomica palermitana', Prot. 646/14/cnc/fta. Consiglio Nazionale dei Chimici (CNC), Rome. Available https://www.chimicifisici.it/wp-con tent/uploads/2018/10/ARANCINA_CNC__STG_2014.pdf. Accessed 8 April 2020

CNC (2014c) DOMANDA DI REGISTRAZIONE - Art. 8 - Regolamento (UE) n. 1151/2012 del Parlamento Europeo e del Consiglio del 21 novembre 2012 sui regimi di qualità dei prodotti agricoli e alimentari - "SFINCIONELLO". Annex to the document 'Accordo tra Comune di Palermo e Consiglio Nazionale dei Chimici (CNC) per la partecipazione alla realizzazione di un sistema di salvaguardia e garanzia della tradizione gastronomica palermitana', Prot. 646/14/cnc/fta. Consiglio Nazionale dei Chimici (CNC), Rome. Available https://www.chimicifisici.it/wp-con tent/uploads/2018/10/SFINCIONELLO_CNC_STG_2014.pdf. Accessed 7 April 2020

CNC (2014d) DOMANDA DI REGISTRAZIONE - Art. 8 - Regolamento (UE) n. 1151/2012 del Parlamento Europeo e del Consiglio del 21 novembre 2012 sui regimi di qualità dei prodotti agri-coli e alimentari - "PANE CA MEUSA". Annex to the document 'Accordo tra Comune di Palermo e Consiglio Nazionale dei Chimici (CNC) per la partecipazione alla realizzazione di un sistema di salvaguardia e garanzia della tradizione gastronomica palermitana', Prot. 646/14/cnc/fta. Consiglio Nazionale dei Chimici (CNC), Rome. Available https://www.chimicifisici.it/wp-con tent/uploads/2018/10/PANE_CA_MEUSA_rev.3dic2014.pdf. Accessed 8 April 2020

CNC (2014e) DOMANDA DI REGISTRAZIONE DI UNA STG - Art. 8 - Regolamento (UE) n. 1151/2012 del Parlamento Europeo e del Consiglio del 21 novembre 2012 sui regimi di qualità dei prodotti agricoli e alimentary - "PANE E PANELLE". Annex to the document 'Accordo tra Comune di Palermo e Consiglio Nazionale dei Chimici (CNC) per la partecipazione alla realiz-zazione di un sistema di salvaguardia e garanzia della tradizione gastronomica palermitana', Prot. 646/14/cnc/fta. Consiglio Nazionale dei Chimici (CNC), Rome.Available https://www.chimic ifisici.it/wp-content/uploads/2018/10/PANE_E_PANELLE__CNC_STG_2014.pdf. Accessed 7 April 2020

Cocks P, Siddique K, Hanbury C (2000) Lathyrus: a new legume. A Report for the Rural Indus-tries Research and Development Corporation, Jan 2000, RIRDC Publication No 99/150, RIRDC Project No UWA-21A. Rural Industries Research & Development Corporation (RIRDC), now AgriFutures Australia, Wagga Wagga

Cortese RDM, Veiros MB, Feldman C, Cavalli SB (2016) Food safety and hygiene practices of vendors during the chain of street food production in Florianopolis, Brazil: A cross-sectional study. Food Control 62:178–186. https://doi.org/10.1016/j.foodcont.2015.10.027

Costantini L, Costantini L, Napolitano G, Whitehouse D (1983) Cereali e legumi medievali provenienti dalle mura di Santo Stefano, Anguillara Sabazia (Roma). Archeologia Medievale 10:393–413

Dauverd C (2006) Genoese and Catalans: trade diaspora in Early Modern Sicily. Mediterr Stud 15:42–61

De Castro J (1958) Science and mankind. No. 1. Hunger and food. World Federation of Scientific Workers, London

De Haen H, Réquillart V (2014) Linkages between sustainable consumption and sustainable produc-tion: some suggestions for foresight work. Food Secur 6(1):87–100. https://doi.org/10.1007/s12 571-013-0323-3

de Suremain CÉ (2016) The never-ending reinvention of 'traditional food'. In: Sébastia B (ed) Eating traditional food: politics, identity and practices. Routledge, Abingdon

Delgado AM, Almeida MDV, Parisi S (2017) Chemistry of the Mediterranean diet. Springer International Publishing, Cham. https://doi.org/10.1007/978-3-319-29370-7

Deshpande SS (1992) Food legumes in human nutrition: a personal perspective. Crit Rev Food Sci Nutr 32(4):333–363. https://doi.org/10.1080/10408399209527603

Deshpande SS, Campbell CG (1992) Genotype variation in BOAA, condensed tannins, phenolics and enzyme inhibitors of grass pea (*Lathyrus sativus*). Can J Plant Sci 72(4):1037–1047. https://doi.org/10.4141/cjps92-130

D'Silva J, Webster J (eds) (2017) The meat crisis: Developing more sustainable and ethical production and consumption. Routledge, London

Duke JA, Reed CF, Weder JKP (1981) Lathyrus sativus L. In: Duke JA (ed) Handbook of legumes of world economic importance. Plenum Press, New York and London, pp 107–110

Ebert AW (2014) Potential of underutilized traditional vegetables and legume crops to contribute to food and nutritional security, income and more sustainable production systems. Sustain 6(1):319–335. https://doi.org/10.3390/su6010319

Elia M (2000) Hunger disease. Clin Nutr 19(6):379–386. https://doi.org/10.1054/clnu.2000.0157

Esclassan R, Grimoud AM, Ruas MP, Donat R, Sevin A, Astie F, Lucas S, Crubezy E (2009) Dental caries, tooth wear and diet in an adult medieval (12th–14th century) population from Mediterranean France. Arch Oral Biol 54(3):287–297. https://doi.org/10.1016/j.archoralbio. 2008.11.004

Falkenmark M (2013) Growing water scarcity in agriculture: future challenge to global water security. Philos Trans Royal Soc A Math Phys Eng Sci 371(2002):20120410. https://doi.org/10. 1098/rsta.2012.0410

FAO (2015) Classification of Commodities (draft). 4. Pulses and Derived Products. Food and Agriculture Organization of the United Nations (FAO), Rome. Available http://www.fao.org/es/fao def/fdef04e.htm. Accessed 10 July 2020

Fernández-Armesto F (2002) Near a thousand tables: a history of food. Free Press, Simon and Schuster, New York

Fogel RW (2004) The escape from hunger and premature death, 1700–2100: Europe, America, and the Third World, vol 38. Cambridge University Press, Cambridge

Gerbens-Leenes PW, Nonhebel S, Krol MS (2010) Food consumption patterns and economic growth. Increasing affluence and the use of natural resources. Appetite 55(3):597–608. https://doi.org/10.1016/j.appet.2010.09.013

Giovannucci E, Rimm EB, Stampfer MJ, Colditz GA, Ascherio A, Willett WC (1994) Intake of fat, meat, and fiber in relation to risk of colon cancer in men. Cancer Res 54(9):2390–2397

Gresta F, Rocco C, Lombardo GM, Avola G, Ruberto G (2014) Agronomic characterization and α-and β-ODAP determination through the adoption of new analytical strategies (HPLC-ELSD and NMR) of ten Sicilian accessions of grass pea. J Agric Food Chem 62(11):2436–2442. https://doi.org/10.1021/jf500149n

Guigoni A (2004) La cucina di strada Con una breve etnografia dello street food genovese. Mneme-Revista de Humanidades 4(9 fev./mar. de 2004):32–43

Hanbury CD, Siddique KHM, Galwey NW, Cocks PS (1999) Genotype-environment interaction for seed yield and ODAP concentration of Lathyrus sativus L. and L. cicera L. in Mediterranean-type environments. Euphytica 110(1):45–60. https://doi.org/10.1023/a:1003770216955

Heinzelmann U (2014) Beyond bratwurst: a history of food in Germany. Reaktion Books Ltd, London

Helou A (2006) Mediterranean street food: stories, soups, snacks, sandwiches, barbecues, sweets, and more from Europe, North Africa, and the Middle East. Harper Collins Publishers, New York

Hillocks RJ, Maruthi MN (2012) Grass pea (Lathyrus sativus): is there a case for further crop improvement? Euphytica 186(3):647–654. https://doi.org/10.1007/s10681-012-0702-4

Hoekstra AY (2012) The hidden water resource use behind meat and dairy. Anim Front 2(2):3–8. https://doi.org/10.2527/af.2012-0038

Jiménez-Colmenero F (2007) Healthier lipid formulation approaches in meat-based functional foods. Technological options for replacement of meat fats by non-meat fats. Trends Food Sci Technol 18(11):567–578. https://doi.org/10.1016/j.tifs.2007.05.006

Kaplan L (2008) Legumes in the history of human nutrition. In: Du Bois CM, Tan CB, Mintz SW (eds) The World of Soy. University of Illinois Press, Champaign, pp 27–44

Kennedy G, Nantel G, Shetty P (2003) The scourge of hidden hunger: global dimensions of micronutrient malnutrition. Food Nutr Agric 32:8–16

Kislev ME (1989) Origins of the cultivation of Lathyrus sativus and L. cicera (fabaceae). Econ Botan 43(2):262–270. https://doi.org/10.1007/BF02859868

Kollnig S (2020) The 'good people' of Cochabamba city: ethnicity and race in Bolivian middle-class food culture. Lat Am Caribb Ethn Stud 15(1):23–43. https://doi.org/10.1080/17442222.2020.169 1795

Kumar J, Pratap A, Solanki RK, Gupta DS, Goyal A, Chaturvedi SK, Nadarajan N, Kumar S (2011a) Genomic resources for improving food legume crops. J Agric Sci (Cambridge) 150:289–318. https://doi.org/10.1017/S0021859611000554

Kumar S, Bejiga G, Ahmed S, Nakkoul H, Sarker A (2011b) Genetic improvement of grasspea for low neurotoxin (B-ODAP) content. Food Cheml Toxicol 49(3):589–600. https://doi.org/10.1016/j.fct.2010.06.051

Kumar S, Gupta P, Barpete S, Sarker A, Amri A, Mathur PN, Baum M (2013) Grass pea. In: Singh M, Upadhyaya HD, Singh Bisht (eds) Genetic and genomic resources of grain legume improvement. Elsevier, London and Walthan, pp 269–292. https://doi.org/10.1016/b978-0-12-397935-3.00011-6

Lambein F, Diasolua Ngudi D, Kuo YH (2010) Progress in prevention of toxico-nutritional neurodegenerations. Afr Technol Develop Forum J 6(3–4):60–65

Lambein F, Kuo YH (1997) Lathyrus sativus, a neolithic crop with a modern future?: an overview of the present situation. In: Proceedings of the international conference 'Lathyrus sativus—cultivation and nutritional value in animals and human', Lublin, Radom, Poland, 9–10 June 1997, pp 6–12

Larcher C, Camerer S (2015) Street food. Temes de Dissen 31:70–83. Available https://core.ac.uk/download/pdf/39016176.pdf. Accessed 9 April 2020

Le Bras-Goude G (2011) Reconstructing past populations' behaviors—diet, Bones and isotopes in the Mediterranean. TÜBA J Archaeol (TÜBA-AR) 14: 215–229

Legume Phylogeny Working Group, Bruneau A, Doyle JJ, Herendeen P, Hughes CE, Kenicer G, Lewis G, Mackinder B, Pennington RT, Sanderson MJ, Wojciechowski MF, Koenen E (2013) Legume phylogeny and classification in the 21st century: progress, prospects and lessons for other species-rich clades. Taxon 62(2):217–248. https://doi.org/10.12705/622.8

Ljuština M, Mikić A (2008) Grain legumes technology transfer in Old Europe-archaecological evidence. In: Proceedings of grain legumes technology transfer platform (GL-TTP) Workshop, 2, Novi Sad (Serbia), 27–28 November 2008. Institute of Field and Vegetable Crops, Novi Sad

Macas J, Novak P, Pellicer J, Čížková J, Koblížková A, Neumann P, Fuková I, Doležel J, Kelly LJ, Leitch IJ (2015) In depth characterization of repetitive DNA in 23 plant genomes reveals sources of genome size variation in the legume tribe *Fabeae*. PLoS ONE 10(11):e0143424. https://doi.org/10.1371/journal.pone.0143424

Malysheva SV, Larionova DA, Diana Di Mavungu J, De Saeger S (2014) Pattern and distribution of ergot alkaloids in cereals and cereal products from European countries. World Mycotoxin J 7(2):217–230. https://doi.org/10.3920/WMJ2013.1642

Manning JA (2009) Constantly containing. Dissertation, West Virginia University

Maranzano B (2014) Lo sviluppo del fenomeno "street food": il cibo di strada a Palermo ieri e oggi. Dissertation, University of Pisa, Italy

Mayer J (1976) The dimensions of human hunger. Sci Am 235(3):40–49

McHugh MR (2015) Modern Palermitan markets and street food in the Ancient Roman World. Conference paper, the Oxford Symposium on Food and Cookery, St. Catherine's College, Oxford University

Mekonnen MM, Hoekstra AY (2012) A global assessment of the water footprint of farm animal products. Ecosyst 15(3):401–415. https://doi.org/10.1007/s10021-011-9517-8

Metrick A, Weitzman ML (1998) Conflicts and choices in biodiversity preservation. J Econ Perspect 12(3):21–34. https://doi.org/10.1257/jep.12.3.21

Mikić A, Mihailović V, Ćupina B, Đurić B, Krstić Đ, Vasić M, Vasiljević S, Karagić V, Đorđević V (2011) Towards the re-introduction of grass pea (*Lathyrus sativus*) in the West Balkan Countries: the case of Serbia and Srpska (Bosnia and Herzegovina). Food Chem Toxicol 49(3):650–654. https://doi.org/10.1016/j.fct.2010.07.052

Minamioka H, Imai H (2009) Sphingoid long-chain base composition of glucosylceramides in Fabaceae: a phylogenetic interpretation of Fabeae. J Plant Res 122(4):415. https://doi.org/10.1007/s10265-009-0227-7

Molotoks A, Kuhnert M, Dawson TP, Smith P (2017) Global hotspots of conflict risk between food security and biodiversity conservation. Land 6(4):67. https://doi.org/10.3390/land6040067

Montanari M (2001) Cucina povera, cucina ricca. Quaderni medievali 52:95–105

Morton PE (2014) Tortillas: a cultural history. University of New Mexico Press, Albuquerque

Murti VVS, Seshadri TR, Venkitsubramanian TA (1964) Neurotoxic compounds of the seeds of *Lathyrus sativus*. Phytochem 3(1):73–78. https://doi.org/10.1016/S0031-9422(00)83997-7

Nene YL (2006) Indian pulses through the millennia. Asian Agri-History 10(3):179–202. Available https://www.asianagrihistory.org/pdf/volume10/Indian_pulses.pdf. Accessed 10 July 2020

Norton GW, Alwang J, Masters WA (2009) Economics of agricultural development, 2nd edn. Routledge, London and New York

Nzwalo H, Cliff J (2011) Konzo: from poverty, cassava, and cyanogen intake to toxico-nutritional neurological disease. PloS Negl Trop Dis 5(6):e1051. https://doi.org/10.1371/journal.pntd.000 1051

Ochatt S, Conreux C, Smýkalová I, Smýkal P, Mikić A (2016) Developing biotechnology tools for 'beautiful' vavilovia (*Vavilovia formosa*), a legume crop wild relative with taxonomic and agronomic potential. Plant Cell, Tissue Organ Cult 127(3):637–648. https://doi.org/10.1007/s11 240-016-1133-z

Osemeobo GJ (1993) Impact of land use on biodiversity preservation in Nigerian natural ecosystems: a review. Nat Res J 33(4):1015–1025

Oskoueiyan R, Kazempour Osaloo S, Amirahmadi A (2014) Molecular phylogeny of the genus lathyrus (Fabaceae-Fabeae) based on cpDNA matK sequence in Iran. Iran J Biotechnol 12(2):41–48. https://doi.org/10.5812/IJB.10315

Patel K, Guenther D, Wiebe K, Seburn RA (2014) Promoting food security and livelihoods for urban poor through the informal sector: a case study of street food vendors in Madurai, Tamil Nadu, India. Food Sec 6(6):861–878. https://doi.org/10.1007/s12571-014-0391-z

Peh KS (2011) Crop failure signals biodiversity crisis. Nature 473(7347):284. https://doi.org/10.1038/473284d

Pilcher JM (2017) Food in world history, 2nd edn. Routledge, Taylor & Francis Group, New York and London

Pilcher JM (2017b) Planet taco: a global history of Mexican food. Oxford University Press, Oxford

Pisano A (2011) La farinata diventa fainé. Un esempio di indigenizzazione. Intrecci. Quaderni di antropologia cultural I, 1:35–58. Associazione Culturale Demo Etno Antropologica, Sassari

Popkin BM, Horton S, Kim S, Mahal A, Shuigao J (2001) Trends in diet, nutritional status, and diet-related noncommunicable diseases in China and India: the economic costs of the nutrition transition. Nutr Rev 59(12):379–390. https://doi.org/10.1111/j.1753-4887.2001.tb06967.x

Rao SLN, Adiga PR, Sarma PS (1964) Rochon JJ, Doyle CJ, Greef JM, Hopkins A, Molle G, Sitzia M, Scholefield D, Smith CJ (2004) Grazing legumes in Europe: a review of their status, management, benefits, research needs and future prospects. Grass Forage Sci 59(3):197–214. https://doi.org/10.1111/j.1365-2494.2004.00423.x

Saraswat KS (1980) The ancient remains of the crop plants at Atranjikhera (c. 2000–1500 BC). J Indian Botan Soc 59(4):306–319

Sathe SK (1996) The nutritional value of selected Asiatic pulses: chickpea, black gram, mung bean and pigeon pea. In: Nwokolo E, Smartt J (eds) Food and feed from legumes and oilseeds. Springer, Boston, pp 12–32. https://doi.org/10.1007/978-1-4613-0433-3_2

Sawicka D, Brzezińska J, Bednarczyk M (2011) Cryoconservation of embryonic cells and gametes as a poultry biodiversity preservation method. Folia Biol 59(1–2):1–5. https://doi.org/10.3409/fb59_1-2.01-05

Schaefer H, Hechenleitner P, Santos-Guerra A, de Sequeira MM, Pennington RT, Kenicer G, Carine MA (2012) Systematics, biogeography, and character evolution of the legume tribe Fabeae with special focus on the middle-Atlantic island lineages. BMC Evol Biol 12(1):250. https://doi.org/10.1186/1471-2148-12-250

Schuster-Gajzágó I (2004) Nutritional aspects of legumes. Cultivated Plants, Primarily as Food Sources 1:1–7. Encyclopedia of Life Support Systems (EOLSS), UNESCO-EOLSS Joint Committee, Paris. Available http://www.eolss.net/sample-chapters/c10/E5-02-02.pdf. Accessed 9 July 2020

Sharma S, Bhagat A, Parisi S (2019) Raw material scarcity and overproduction in the food industry. Springer International Publishing, Cham. https://doi.org/10.1007/978-3-030-14651-1

Sheen B (2010) Foods of Egypt. Greenhaven Publishing LLC, New York

Shiva V (2016) Water wars: Privatization, pollution, and profit. South End Press, Cambridge, MA

Simopoulus AP, Bhat RV (2000) Street foods. Karger AG, Basel

Sinou C, Cardinal-McTeague W, Bruneau A (2020) Testing generic limits in Cercidoideae (Leguminosae): insights from plastid and duplicated nuclear gene sequences. Taxon 69(1):67–86. https://doi.org/10.1002/tax.12207

Sozer N, Holopainen-Mantila U, Poutanen K (2017) Traditional and new food uses of pulses. Cereal Chem 94(1):66–73. https://doi.org/10.1094/CCHEM-04-16-0082-FI

Spash CL, Hanley N (1995) Preferences, information and biodiversity preservation. Ecolog Econ 12(3):191–208. Available https://mpra.ub.uni-muenchen.de/38351/. Accessed 8 July 2020

Spencer PS, Palmer VS (2003) Lathyrism: aqueous leaching reduces grass-pea neurotoxicity. Lancet 362(9398):1775–1776

Sraïri MT, Rjafallah M, Kuper M, Le Gal PY (2009) Water productivity through dual purpose (milk and meat) herds in the Tadla irrigation scheme. Morocco. Irrig Drain 58(S3):S334–S345. https://doi.org/10.1002/ird.531

Steven QA (2018) Fast food, street food: Western fast food's influence on fast service food in China. Dissertation, Duke University, Durham

Stoddard FL, Hovinen S, Kontturi M, Lindström K, Nykänen A (2009) Legumes in Finnish agriculture: history, present status and future prospects. Agric Food Sci 18:191–205

Sunderland TC (2011) Food security: why is biodiversity important? Int For Rev 13(3):265–274. https://doi.org/10.1505/146554811798293908

Tadesse W, Bekele E (2003) Variation and association of morphological and biochemical characters in grass pea (*Lathyrus sativus* L.). Euphytica 130(3):315–324. https://doi.org/10.1023/a:1023087903679

Thang NM, Popkin BM (2004) Patterns of food consumption in Vietnam: effects on socioeconomic groups during an era of economic growth. Eur J Clin Nutr 58(1):145–153. https://doi.org/10.1038/sj.ejcn.1601761

The Global Pulse Confederation (2020) What are pulses? The Global Pulse Confederation, Dubai. https://pulses.org/what-are-pulses

Toussaint-Samat M (2009) A history of food. Wiley, Chichester

Trichopoulou A, Lagiou P (1997) Healthy traditional Mediterranean diet: an expression of culture, history, and lifestyle. Nutr Rev 55(11):383–389. https://doi.org/10.1111/j.1753-4887.1997.tb01578.x

Tscharntke T, Clough Y, Wanger TC, Jackson L, Motzke I, Perfecto I, Vandermeer J, Whitbread A (2012) Global food security, biodiversity conservation and the future of agricultural intensification. Biol Conserv 151(1):53–59. https://doi.org/10.1016/j.biocon.2012.01.068

Turrell G, Kavanagh AM (2006) Socio-economic pathways to diet: modelling the association between socio-economic position and food purchasing behaviour. Pub Health Nutr 9(3):375–383. https://doi.org/10.1079/PHN2005850

Urga K, Debella A, Agata N, Bayu A, Zewdie W (2002) Laboratory studies on the outbreak of gangrenous ergotism associated with consumption of contaminated barley in Arsi, Ethiopia. Ethiop J Health Dev 16(3):317–323. https://doi.org/10.4314/ejhd.v16i3.9800

Urga K, Fite A, Kebede B (1995) Nutritional and anti-nutritional factors of grass pea germ plasm. Bull Chem Soc Ethiop 9:9–16

van der Goot AJ, Pelgrom PJ, Berghout JA, Geerts ME, Jankowiak L, Hardt NA, Boom RM (2016) Concepts for further sustainable production of foods. J Food Eng 168:42–51

van Keulen H (1975) Simulation of water use and herbage growth in arid regions. Dissertation, Centre for Agricultural Publishing and Documentation, Wageningen

Vernon J (2007) Hunger: a modern history. Harvard University Press, Harvard

Webb RE, Hyatt SA (1988) Haitian street foods and their nutritional contribution to dietary intake. Ecol Food Nutr 21(3):199–209. https://doi.org/10.1080/03670244.1988.9991033

Westgren RE (1999) Delivering food safety, food quality, and sustainable production practices: the Label Rouge poultry system in France. Am J Agric Econ 81(5):1107–1111. https://doi.org/10.2307/1244092

Wilkins J, Hill S (2009) Food in the ancient world. Blackwell Publishing Ltd, Maiden, Oxford, and Carlton

Wilkinson J (2004) The food processing industry, globalization and developing countries. Electron J Agric Develop Econ 1(2):184–201

Williamson CS, Foster RK, Stanner SA, Buttriss JL (2005) Red meat in the diet. Nutr Bull 30(4):323–355

Wittman H, Chappell MJ, Abson DJ, Kerr RB, Blesh J, Hanspach J, Perfecto I, Fischer J (2017) A social–ecological perspective on harmonizing food security and biodiversity conservation. Reg Environ Change 17(5):1291–1301. https://doi.org/10.1007/s10113-016-1045-9

Chapter 2
Lathyrus sativus. Traditional Grass Pea-Based Foods

Abstract This Chapter concerns the traditional use of a peculiar crop legume, grass pea (*Lathyrus sativus*) as a food product or ingredient for typical food products. At present, this legume is known worldwide and widely used in relation to human nutrition. In fact, grass pea has very interesting nutritional features, including the abundance in proteins, and peculiar organoleptic properties. On the other hand, grass pea may be questioned with concern to human and animal nutrition because of a peculiar neurotoxin associated with neurolathyrism and occurs if grass pea-based foods are consumed in notable quantities. However, these food products are an interesting legacy for many geographical and ethnic cultures. At present, many food varieties are reported to be correlated with grass pea worldwide. An overview of different food versions ascribed to *L. sativus* in the world is offered here in relation to continental areas and taking into mind the existence of a peculiar culinary pattern in Ethiopia (soups/sauces/processed seeds/unleavened breads) which can be partially found in other non-African nations.

Keywords Cereal · Grain legume · Grass pea · Processed seed · Sauce · Soup · Unleavened bread

2.1 Grass Pea and Associated Recipes Worldwide. An Introduction

The use of grass pea (*Lathyrus sativus*) in human nutrition is well-documented worldwide, in spite of the unfavourable reputation for this legume because of the 'neurolathyrism' disease associated with excessive consumption of this food (Chap. 1). This premise has solid-based because of the notable number of food recipes associated with grass pea, alone or in association with cereals or other grain legumes such as chickpea (*Cicer arietinum*). Some historical Italian recipes are based on cereals and/or grain legumes (Abulafia 1978; Barbagli and Barzini 2010; Carrabino 2017; CNC 2014a, b, c, d, e; Dauverd 2006; Guigoni 2004; Pisano 2011).

© The Author(s), under exclusive license to Springer
Nature Switzerland AG 2020
M. Barone and R. Tulumello, *Lathyrus sativus and Nutrition*,
Chemistry of Foods, https://doi.org/10.1007/978-3-030-59091-8_2

   With reference to agricultural activities, grass pea is able to survive in extremely difficult environmental conditions, including excessive heat, water stagnation, famine, and drought (Ben-Brahim et al. 2002; Campbell 1997; Cocks et al. 2000; Kumar et al. 2011a, b, 2013). Apart the limited and recent introduction of this grain legume (and also *L. cicera,* known as chickling vetch) in Australia (because of the need of pulses, in particular when speaking of fine-textured soils with pH ranging between extreme low and high values), grass pea is observed in the North and the Sub-Saharan area of Africa, in South Asia countries, Southern Europe, and South America also (Cocks et al. 2000). Actually, *L. sativus* is important when speaking of human and animal supplementation. In relation to human nutrition only, grass pea is cited in many countries and areas including Bangladesh, Ethiopia, India, Nepal, Pakistan, Bulgaria, Germany, Greece, Italy, Poland, Portugal, Spain, Portugal, Crete, Southern Russia, Rhodes, Cyprus, USA, Chile, Syria, Turkey, and the Maghreb (Campbell 1997; Kumar et al. 2013). In general, *Lathyrus* genus seems to be a constant presence in the human history (Fig. 2.1) and in association with grain legumes, as the signal of a cultural heritage (Alfiero et al. 2017; Bell et al. 2014; Booth and Coveney

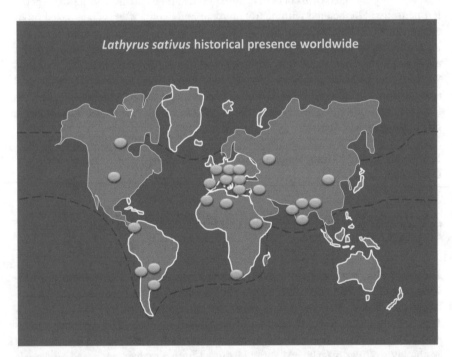

Fig. 2.1 In relation to human nutrition only, grass pea is cited in many countries (Campbell 1997; Kumar et al. 2013). In general, *Lathyrus* genus seems to be a constant presence in the human history and in association with grain legumes. Interestingly, the diffusion area does not appear to include regions with extreme environmental conditions (outside the two red-dashed lines and the central area). However, grass pea is highly resistant to drastic conditions. It has to be considered that the recent introduction projects concerning *L. sativus* in Australia are not mentioned here because the picture should only consider the historical influence and presence of grass pea in the human history.

2007; Chong and Eun 1992; Cortese et al. 2016; de Suremain 2016; Delgado et al. 2017; Heinzelmann 2014; Helou 2006; Kollnig 2020; Larcher and Camerer 2015; Manning 2009; Maranzano 2014; McHugh 2015; Morton 2014; Patel et al. 2014; Pilcher 2017; Sheen 2010; Simopoulos and Bhat 2000; Steven 2018; Webb and Hyatt 1988; Wilkins and Hill 2009; Wilkinson 2004).

Maybe, one of problems associated with *L. sativus* (excluding the safety and health concerns under the 'neurolathyrism' name) might be the notable number of similar and different 'common names' for this legume (Chap. 1). As a simple example, the common name 'chickling vetch' is ascribed both to *L. sativus* and *L. cicera*, and in India at least both species have the same name (Campbell 1997)... Consequently, the same legume may be differently found in different countries, and this circumstance may explain partially the confusion concerning *L. sativus* and recipes containing this food.

2.2 Grass Pea and Associated Recipes. A Continental Analysis

In general, grass pea is used worldwide as supplementation for cereal-based foods with the aim of enhancing nutritional properties of cereal-based recipes. It has been reported that cereal mixtures composed of maize (*Zea mays*) and wheat (*Triticum aestivum*) can contain intentionally grass pea with the aim of enhancing calories and vegetable fat matter in the final meal, in Ethiopia. These preparations—boiled, fermented pancake, unleavened bread—can be dangerous enough, and even its spastic paraparesis episodes may be less frequent if more than three cereals are used in the preparation. Alternatively, roasted *L. sativus* and unripened green seeds may be used as the first ingredient (with some condiments), as a traditional recipe (Getahun et al. 2003).

As a consequence, Ethiopia can serve as a useful example when speaking of grass pea-based preparations worldwide. Also, this case can be used in relation to African countries.

2.2.1 Grass Pea and Associated Recipes. Africa

L. sativus is extremely used in Ethiopia and Eritrea, and the main grass pea-based preparations of interest can be summarised as follows (Brink and Belay 2006; Urga et al. 1995–2005):

(a) A spiced soup-like food, the traditional *shiro wot* or *shiro wott*
(b) A sauce containing dehusked split seeds: its name is *kik wot*

(c) A peculiar food made of roasted whole seeds (the traditional *kollo*) or roasted whole seeds (the *nifro*)
(d) An unleavened flat bread named *kitta*.

In addition, it is reported that legume flours may be adulterated with grass pea flour. In this ambit, adulterated flours are chickpea or dry peas (Urga et al. 1995–2005). Grass pea flours are reported to be used as fraudulent additives for legume flours such as chickpeas, dry peas, and Bengal gram. This adulteration is extremely worrying because of the high rate of substitution on the one hand and obvious health and safety problems related to the unknown *L. sativus* presence (heat treatments can be used but without detailed information) (Teklehaimanot et al. 1993).

With the exclusion of food fraud possibilities, it appears that Ethiopians prefer to use grass pea in four different versions:

(a) Soups (or soup-like foods)
(b) Sauces
(c) Processed seeds
(d) Unleavened breads.

It has to be considered that above-mentioned products can be made (with the exclusion of whole seeds) using more than one single raw material. It has been reported that grass pea could be used to lower the toxicity of cassava-based foods. Cassava (*Manihot esculenta*) contains free and bound linamarin and lotaustralin (two cyanogenic glucosides) and a peculiar enzyme (linamarase) able to react with these molecules. The result is the production of hydrogen cyanide. For this reason, the toxicity of cassava-based foods could be lowered with maize, other cereals, and also grass pea (Kebede et al. 2012).

Could the same 'culinary pattern' be found in other African countries and areas? The answer is probably positive and surely demonstrated by means of archaeological studies. As a simple example, *L. sativus* residues have been found as royal funeral offerings in ancient Egyptian eras (Lambein et al. 2010). However, a certain lack of studies concerning grass pea-based foods in Africa appears at present, with the important exclusion of Ethiopian recipes (and some experimental studies concerning cereal-based foods), in our knowledge.

However, the Ethiopian 'model' could be replicated in other continental areas, when speaking of anthropic traditions and related food uses.

2.2.2 Grass Pea and Associated Recipes. Asia

In relation to Asian countries, a good example is Bangladesh. It is reported that *L. sativus* (*khesari*) split seeds are used as component of soups. Also, grass pea seeds (*besan*) can be turned into flours or powders for breads and local preparations named *pakoras, chapati*, or *dalpuri*. In addition, grass pea can be consumed alone as leafy vegetable (Arora et al. 1996). It should be considered that this grain legume is associated with rice farming economies in the West Bengal area, and consequently grass pea is not perceived as an extraneous product (Das 2000).

In summary, this 'pattern' (also recognised in other Asian countries such as India, Nepal, and Pakistan) is similar to the Ethiopian situation. In these countries, grass pea is a legume for poor people; unfortunately, this anthropic classification may be correlated with neurolathyrism fatal episodes.

2.2.3 Grass Pea and Associated Recipes. Oceania and the Americas

In relation to American countries, there are not important information concerning grass pea and current uses for human nutrition, while it may be used as green manure in Canada at least (Brink and Belay 2006). In addition, grass pea is cultivated in the USA as a fodder crop (McCutchan 2003). This result is naturally dependent on the lathyrism danger. Consequently, the presence of *L. sativus* is confirmed both in North and South America without connections to human nutrition (Gutiérrez et al. 2001). For these reasons, there are not particular food recipes containing grass pea in our knowledge when speaking of the Americas. This result may appear strange enough because of the above-discussed 'pattern' which associates grass pea with cereals (and Central America is well known for the abundant cultivation of maize). In addition, the name for grass pea in the America is reported to be 'Spanish field pea' (Shurtleff and Aoyagi 2009), maybe in connection with one of the most important original areas, the Mediterranean basin.

On the other hand, it has to be remembered that grass pea is important when speaking of (1) agricultural resistance to adverse environmental conditions and (2) occurrence of famines. For these reasons, the use of grass pea is important in other world areas such as West Bengal or Equatorial Africa. The same thing can be confirmed when speaking of Australia, where grass pea has been introduced very recently. In this country also, *L. sativus* is apparently interesting as a green manure, at present (McCutchan 2003). Consequently, there are not anthropic traditions linked with *L. sativus* in Oceania.

2.2.4 Grass Pea and Associated Recipes. Europe

The cultivation of grain legumes—lentils, peas, and bitter vetches—spread in the Neolithic Europe since 7500 B.C. at least (Santa Maira, Iberian Peninsula; Nevalı Çori, Turkey), while *Faba* legumes became the most used legume after some time in some sites. As discussed in Chap. 1, there is a certain connection between the diffusion of legumes in ancient European civilisations and the use of fire in food preparation. With reference to grass pea, the possible occurrence of digestion problems by early hominids and the concomitant production of some neurotoxins and trypsin inhibitors could have been solved with a simple solution: chemical denaturation by means of cooking treatments.

As mentioned above, could the Ethiopian 'culinary pattern' be found in Europe? This pattern concerns the use of grass pea in four different versions:

(a) Soups (or soup-like foods)
(b) Sauces
(c) Processed seeds
(d) Unleavened breads.

It appears that this pattern is recognisable in old European traditions. In fact, archaeological studies in Greek sites have demonstrated that the use of grass pea is known since the Bronze Age at least, while the toxicity problem was well identified since Hippocrates' times. Interestingly, modern Greeks are accustomed to eat grass pea similarly to their ancestors: split seeds are boiled in water and subsequently cooked with the early removal of 'heads' (Valamoti 2009). It has to be noted that this tradition, common to Spain and also Ethiopia, corresponds to the third 'pattern' choice: 'processed seeds'.

With concern to soups or similar dishes, grass pea is reported to be soaked and eaten in forms of stews or turned in flour for the preparation of traditional (*gachas*) in Spain at least (AECOSAN 2018; Valamoti 2009). This use is reported in ancient religious traditions when speaking of prepared grass pea stews (Peña-Chocarro and Peña 1999). The production of different gruels, very common in other European areas such as Central and Northern Italy (when *Cicer arietinum* flours are used for chickpea flat breads), demonstrates the large use of grass pea flours since the High Middle Age when speaking of breads from different flours and popular soups (Cherubini 2000; Città di Troina 2017; Costantini et al. 1983; Montanari 2001). With concern to breads, the use of *L. sativus* flour has been suggested recently in an experimental work concerning grass pea Polish cultivars. In addition, grass pea has been suggested in Portugal as a partial soybean replacement for the traditional Japanese *miso* (a fermented food), demonstrating the strong interest for the use of this grain legume even in non-traditionally European recipes. Apparently, only one of the 'Ethiopian culinary pattern' typologies is not found in European countries: the use of grass pea for sauces. In our knowledge, these preparations are not typically found in Europe.

In general, grass pea is now found in Bulgaria, Germany, Greece, Italy, Poland, Portugal, Spain, Portugal, Crete, southern Russia, Rhodes, and Cyprus, when speaking of the European area. A peculiar case study which can represent and explain the historical and nutritional importance of grass pea is discussed in Chap. 3 when speaking of Italian *cicerchia* recipes (*Ru muocche, Cicerchia di Serra de' Conti, Vellutata di cicerchie*, etc.) are also mentioned. Interestingly, all of these recipes are known in the current form since the Middle Age at least.

References

Abulafia D (1978) Pisan commercial colonies and consulates in twelfth-century Sicily. Eng Historical Rev 93(366):68–81

AECOSAN (2018) Report of the Scientific Committee of the Spanish Agency for Consumer Affairs, Food Safety and Nutrition (AECOSAN) on the safety of grass pea flour consumption. Revista del Comité Científico de la AECOSAN 27: 61–78. The Spanish Agency for Consumer Affairs, Food Safety and Nutrition (AECOSAN), Madrid

Akalu G, Johansson G, Nair BM (1998) Effect of processing on the content of β-N-oxalyl-α, β-diaminopropionic acid (gb-ODAP) in grass pea (Lathyrus sativus) seeds and flour as determined by flow injection analysis. Food Chem 62(2):233–237. https://doi.org/10.1016/S0308-814 6(97)00137-4

Alfiero S, Giudice AL, Bonadonna A (2017) Street food and innovation: the food truck phenomenon. Brit Food J 119(11):2462–2476. https://doi.org/10.1108/BFJ-03-2017-0179

Allchin FR (1969) Early cultivated plants in India and Pakistan. In: Ucko J, Dimbleby GW (eds) The domestication and exploitation of plants and animals. Duckworth, London

Arora RK, Mathur PN, Riley KW, Adham Y (1996) Lathyrus genetic resources in Asia. In: Proceedings of a Regional Workshop, 27–29 December 1995, Indira Gandhi Agricultural University, Raipur, India, pp 27–29

Barbagli A, Barzini S (2010) Pane, pizze e focacce. Giunti Editore, Florence

Bell JS, Loukaitou-Sideris A (2014) Sidewalk informality: an examination of street vending regulation in China. Int Plan Stud 19(3–4):221–243. https://doi.org/10.1080/13563475.2014. 880333

Ben-Brahim N, Salho A, Chtorou N, Combes D, Marrakcho M (2002) Isozymic polymorphism and phylogeny of 10 Lathyrus species. Genet Res Crop Evol 49:427–436. https://doi.org/10.1023/A: 1020629829179

Booth SL, Coveney J (2007) Survival on the streets: prosocial and moral behaviors among food insecure homeless youth in Adelaide, South Australia. J Hunger Environ Nutr 2(1):41–53. https://doi.org/10.1080/19320240802080874

Brink M, Belay G (eds) (2006) Plant resources of tropical Africa 1. Cereals and Pulses. PROTA Foundation, Wageningen/Backhuys Publishers, Leiden/CTA, Wageningen

Campbell CG (1997) Grass pea, Lathyrus sativus L. Promoting the conservation and use of underutilized and neglected crops. Institute of Plant Genetics and Crop Plant Research, Rome, and Germany/International Plant Genetic Resources Institute, Gatersleben, Nr 18, Rome, Italy, p 92

Carrabino D (2017) 14 Oratories of the compagnie of Palermo: sacred spaces of rivalry. In: Bullen Presciutti D (ed) Space, place, and motion: locating confraternities in the late medieval and early modern city. Koninklijke Brill Nv (Brill), Leiden, pp 344–371. https://doi.org/10.1163/978900 4339521_016

Cherubini G (2000) L'approvvigionamento alimentare delle città toscane tra il XII e il XIV secolo. Rivista di storia dell'agricoltura 40:33–52

Chong HK, Eun VLN (1992) 4 backlanes as contested regions: construction and control of physical. In: Huat CB, Edwards N (eds) Public space: design, use and management. Singapore University Press, Singapore

Città di Troina (2017) Sottoscritto protocollo d'intesa con il Consiglio Nazionale dei Chimici per la salvaguardia e la garanzia della tradizione gastronomica di Troina. Available https://www.com une.troina.en.it/DETTAGLIO_NEWS.ASP?ID=470. Accessed 14th July 2020

CNC (2014a) Accordo tra Comune di Palermo e Consiglio Nazionale dei Chimici (CNC) per la partecipazione alla realizzazione di un sistema di salvaguardia e garanzia della tradizione gastronomica palermitana, Prot. 646/14/cnc/fta. Consiglio Nazionale dei Chimici (CNC), Rome. Available https://www.chimicifisici.it/wp-content/uploads/2018/10/20131210_accordo_firmato_dal_ Presidente_del_CNC.pdf. Accessed 07 April 2020

CNC (2014b) DOMANDA DI REGISTRAZIONE - Art. 8 - Regolamento (UE) n. 1151/2012 del Parlamento Europeo e del Consiglio del 21 novembre 2012 sui regimi di qualità dei prodotti agricoli e alimentari - " "ARANCINA". Annex to the document 'Accordo tra Comune di Palermo e Consiglio Nazionale dei Chimici (CNC) per la partecipazione alla realizzazione di un sistema di salvaguardia e garanzia della tradizione gastronomica palermitana', Prot. 646/14/cnc/fta. Consiglio Nazionale dei Chimici (CNC), Rome. Available https://www.chimicifisici.it/wp-con tent/uploads/2018/10/ARANCINA_CNC__STG_2014.pdf. Accessed 08 April 2020

CNC (2014c) DOMANDA DI REGISTRAZIONE - Art. 8 - Regolamento (UE) n. 1151/2012 del Parlamento Europeo e del Consiglio del 21 novembre 2012 sui regimi di qualità dei prodotti agricoli e alimentari - "SFINCIONELLO". Annex to the document 'Accordo tra Comune di Palermo e Consiglio Nazionale dei Chimici (CNC) per la partecipazione alla realizzazione di un sistema di salvaguardia e garanzia della tradizione gastronomica palermitana', Prot. 646/14/cnc/fta. Consiglio Nazionale dei Chimici (CNC), Rome. Available https://www.chimicifisici.it/wp-con tent/uploads/2018/10/SFINCIONELLO_CNC_STG_2014.pdf. Accessed 07 Apr 2020

CNC (2014d) DOMANDA DI REGISTRAZIONE - Art. 8 - Regolamento (UE) n. 1151/2012 del Parlamento Europeo e del Consiglio del 21 novembre 2012 sui regimi di qualità dei prodotti agricoli e alimentari - "PANE CA MEUSA". Annex to the document 'Accordo tra Comune di Palermo e Consiglio Nazionale dei Chimici (CNC) per la partecipazione alla realizzazione di un sistema di salvaguardia e garanzia della tradizione gastronomica palermitana', Prot. 646/14/cnc/fta. Consiglio Nazionale dei Chimici (CNC), Rome. Available https://www.chimicifisici.it/wp-con tent/uploads/2018/10/PANE_CA_MEUSA_rev.3dic2014.pdf. Accessed 08 Apr 2020

CNC (2014e) DOMANDA DI REGISTRAZIONE DI UNA STG - Art. 8 - Regolamento (UE) n. 1151/2012 del Parlamento Europeo e del Consiglio del 21 novembre 2012 sui regimi di qualità dei prodotti agricoli e alimentary - "PANE E PANELLE". Annex to the document 'Accordo tra Comune di Palermo e Consiglio Nazionale dei Chimici (CNC) per la partecipazione alla realizzazione di un sistema di salvaguardia e garanzia della tradizione gastronomica palermitana', Prot. 646/14/cnc/fta. Consiglio Nazionale dei Chimici (CNC), Rome.Available https://www.chimicifi sici.it/wp-content/uploads/2018/10/PANE_E_PANELLE__CNC_STG_2014.pdf. Accessed 07 Apr 2020

Cocks P, Siddique K, Hanbury C (2000) Lathyrus: a new legume. A report for the rural industries research and development corporation. January 2000, RIRDC Publication No 99/150, RIRDC Project No UWA-21A. Rural Industries Research & Development Corporation (RIRDC), now AgriFutures Australia, Wagga Wagga

Cortese RDM, Veiros MB, Feldman C, Cavalli SB (2016) Food safety and hygiene practices of vendors during the chain of street food production in Florianopolis, Brazil: a cross-sectional study. Food Control 62:178–186. https://doi.org/10.1016/j.foodcont.2015.10.027

Costantini L, Costantini L, Napolitano G, Whitehouse D (1983) Cereali e legumi medievali provenienti dalle mura di Santo Stefano, Anguillara Sabazia (Roma). Archeologia Medievale 10:393–413

da Costa DMP (2018) Grass pea miso: development of miso based on a portuguese legume-microbiota and preservation capacity. Dissertation, Universidade Nova de Lisboa

Das NR (2000) Lathyrus sativus in rainfed multiple cropping systems in West Bengal, India—a review. Lathyrus Lathyrism Newsl 1:25–27

Dauverd C (2006) Genoese and Catalans: trade diaspora in early modern Sicily. Mediterr Stud 15:42–61

de Suremain CÉ (2016) The never-ending reinvention of 'traditional food'. In: Sébastia B (ed) Eating traditional food: politics, identity and practices. Routledge, Abingdon

Delgado AM, Almeida MDV, Parisi S (2017) Chemistry of the Mediterranean diet. Springer International Publishing, Cham https://doi.org/10.1007/978-3-319-29370-7

Deshpande SS, Campbell CG (1992) Genotype variation in BOAA, condensed tannins, phenolics and enzyme inhibitors of grass pea (Lathyrus sativus). Can J Plant Sci 72(4):1037–1047. https://doi.org/10.4141/cjps92-130

Duke JA, Reed CF, Weder JKP (1981) Lathyrus sativus L. In: Duke JA (ed) Handbook of legumes of world economic importance. Plenum Press, New York and London, pp 107–110

Getahun H, Lambein F, Vanhoorne M, Van der Stuyft P (2003) Food-aid cereals to reduce neuro-lathyrism related to grass-pea preparations during famine. Lancet 362(9398):1808–1810. https://doi.org/10.1016/S0140-6736(03)14902-1

Grela ER, Rybiński W, Klebaniuk R, Matras J (2010) Morphological characteristics of some accessions of grass pea (*Lathyrus sativus* L.) grown in Europe and nutritional traits of their seeds. Genet Res Crop Evol 57(5), 693–701. https://doi.org/10.1007/s10722-009-9505-4

Gresta F, Rocco C, Lombardo GM, Avola G, Ruberto G (2014) Agronomic characterization and α-and β-ODAP determination through the adoption of new analytical strategies (HPLC-ELSD and NMR) of ten Sicilian accessions of grass pea. J Agric Food Chem 62(11):2436–2442. https://doi.org/10.1021/jf500149n

Gryseels G (1986) Difficulties in evaluating on-farm experiments: examples from the Ethiopian highlands. In: Kearl S (ed) Proceedings of a workshop held at the International Livestock Centre for Africa, Addis Ababa, Ethiopia, 24th–27th June 1985, pp 27–56

Guigoni A (2004) La cucina di strada Con una breve etnografia dello street food genovese. Mneme-Revista de Humanidades 4, 9 fev./mar. de 2004:32–43

Gutiérrez JF, Vaquero F, Vences FJ (2001) Genetic mapping of isozyme loci in Lathyrus sativus L. Lathyrus Lathyrism Newsl 2:74–78

Hanbury CD, White CL, Mullan BP, Siddique KHM (2000) A review of the potential of Lathyrus sativus L. and L. cicera L. grain for use as animal feed. Anim Feed Sci Technol 87:1–27. https://doi.org/10.1016/S0377-8401(00)00186-3

Heinzelmann U (2014) Beyond bratwurst: a history of food in Germany. Reaktion Books Ltd, London

Helou A (2006) Mediterranean street food: stories, soups, snacks, sandwiches, barbecues, sweets, and more from Europe, North Africa, and the Middle East. Harper Collins Publishers, New York

Hillocks RJ, Maruthi MN (2012) Grass pea (*Lathyrus sativus*): is there a case for further crop improvement? Euphytica 186(3):647–654. https://doi.org/10.1007/s10681-012-0702-4

Hugon J, Ludolph AC, Spencer PS (2000) β-N-oxalylamino-alanine. In: Spencer PS, Schaumburg H (eds) Experimental and clinical Neurotoxicology, 2nd edn. Oxford University Press, New York, pp 925–938

Kasprzak M, Rzedzicki Z (2012) Application of grasspea wholemeal in the technology of white bread production. Pol J Food Nutr Sci 62(4):207–213. https://doi.org/10.2478/v10222-012-0056-6

Kebede A, Teshome B, Wondimu A, Belay A, Wodajo B, Lakew A (2012) Detoxification and consumption of cassava based foods in South West Ethiopia. Pak J Nutr 11(3):237–242

Kislev ME (1989) Origins of the cultivation of *Lathyrus sativus* and L. cicera (fabaceae). Econ Botan 43(2):262–270. https://doi.org/10.1007/BF02859868

Kollnig S (2020) The 'good people' of Cochabamba city: ethnicity and race in Bolivian middle-class food culture. Lat Am Caribb Ethn Stud 15(1):23–43. https://doi.org/10.1080/17442222.2020.169 1795

Kumar J, Pratap A, Solanki RK, Gupta DS, Goyal A, Chaturvedi SK, Nadarajan N, Kumar S (2011) Genomic resources for improving food legume crops. J Agric Sci (Cambridge) 150:289–318. https://doi.org/10.1017/S0021859611000554

Kumar S, Bejiga G, Ahmed S, Nakkoul H, Sarker A (2011) Genetic improvement of grasspea for low neurotoxin (B-ODAP) content. Food Chem Toxicol 49(3):589–600. https://doi.org/10.1016/j.fct.2010.06.051

Kumar S, Gupta P, Barpete S, Sarker A, Amri A, Mathur PN, Baum M (2013) Grass pea. In: Singh M, Upadhyaya HD, Singh Bisht (eds) Genetic and genomic resources of grain legume improvement. Elsevier, London and Walthan, pp 269–292. https://doi.org/10.1016/B978-0-12-397935-3.00011-6

Lambein F, Diasolua Ngudi D, Kuo YH (2010) Progress in prevention of toxico-nutritional neurodegenerations. Afr Technol Develop Forum J 6(3–4):60–65

Lambein F, Kuo YH (1997) Lathyrus sativus, a neolithic crop with a modern future?: an overview of the present situation. In: Proceedings of the international conference 'Lathyrus sativus—cultivation and nutritional value in animals and human', Lublin, Radom, Poland, 9th –10th June 1997, pp. 6–12

Lambein F, Travella S, Kuo YH, Van Montagu M, Heijde M (2019) Grass pea (Lathyrus sativus L.): orphan crop, nutraceutical or just plain food? Planta 250:821–838. https://doi.org/10.1007/s00425-018-03084-0

Larcher C, Camerer S (2015) Street Food. Temes de Dissen 31:70–83. Available https://core.ac.uk/download/pdf/39016176.pdf. Accessed 09 Apr 2020

Leccese A, Mattonelli S (2008) Pedo-eno-gastronomic Itineraries in Umbria North-eastern Umbria. In: Proceedings of the Convegno internazionale "I paesaggi del vino", Perugia, 8–10th May 2008

Manning JA (2009) Constantly containing. Dissertation, West Virginia University

Maranzano B (2014) Lo sviluppo del fenomeno "street food": il cibo di strada a Palermo ieri e oggi. Dissertation, University of Pisa, Italy

McCutchan JS (2003) A brief history of grasspea and its use in crop improvement. Lathyrus Lathyrism Newsl 3(1):18–23

McHugh MR (2015) Modern Palermitan markets and street food in the Ancient Roman World. Conference paper, the Oxford Symposium on Food and Cookery, St. Catherine's College, Oxford University

Montanari M (2001) Cucina povera, cucina ricca. Quaderni Medievali 52:95–105

Morton PE (2014) Tortillas: a cultural history. University of New Mexico Press, Albuquerque

Murti VVS, Seshadri TR, Venkitsubramanian TA (1964) Neurotoxic compounds of the seeds of Lathyrus sativus. Phytochem 3(1):73–78. https://doi.org/10.1016/S0031-9422(00)83997-7

Nzwalo H, Cliff J (2011) Konzo: from poverty, cassava, and cyanogen intake to toxico-nutritional neurological disease. PloS Negl Trop Dis 5(6):e1051. https://doi.org/10.1371/journal.pntd.0001051

Patel K, Guenther D, Wiebe K, Seburn RA (2014) Promoting food security and livelihoods for urban poor through the informal sector: a case study of street food vendors in Madurai, Tamil Nadu India. Food Sec 6(6):861–878. https://doi.org/10.1007/s12571-014-0391-z

Patto MV, Skiba B, Pang ECK, Ochatt SJ, Lambein F, Rubiales D (2006) Lathyrus improvement for resistance against biotic and abiotic stresses: from classical breeding to marker assisted selection. Euphytica 147(1–2):133. https://doi.org/10.1007/s10681-006-3607-2

Peña-Chocarro L, Peña LZ (1999) History and traditional cultivation of Lathyrus sativus L. and Lathyrus citera L. in the Iberian peninsula. Veget Hist Archaeobotan 8(1–2):49–52. https://doi.org/10.1007/BF02042842

Pilcher JM (2017) Planet taco: a global history of Mexican food. Oxford University Press, Oxford

Pisano A (2011) La farinata diventa fainé. Un esempio di indigenizzazione. Intrecci. Quaderni di antropologia cultural I 1:35–58 (Associazione Culturale Demo Etno Antropologica, Sassari)

Rao SLN, Adiga PR, Sarma PS (1964) The isolation and characterization of β-N-Oxalyl-L-α, β-diaminopropionic acid: a neurotoxin from the seeds of Lathyrus sativus. Biochem 3(3):432–436. https://doi.org/10.1021/bi00891a022

Saraswat KS (1980) The ancient remains of the crop plants at Atranjikhera (c. 2000–1500 BC). J Indian Botan Soc 59(4):306–319

Sheen B (2010) Foods of Egypt. Greenhaven Publishing LLC, New York

Shurtleff W, Aoyagi A (2009) History of soybeans and soyfoods in Mexico and Central America (1877–2009): extensively annotated bibliography and sourcebook. Soyinfo Center, Lafayette

Siddique KHM, Loss SP, Herwig SP, Wilson JM (1996) Growth, yield and neurotoxin (B-ODAP) concentration of three Lathyrus species in Mediterranean type environments of Western Australia. Aust J Exp Agric 36:209–218. https://doi.org/10.1071/EA9960209

Simopoulus AP, Bhat RV (2000) Street foods. Karger AG, Basel

Steven QA (2018) Fast food, street food: Western fast food's influence on fast service food in China. Dissertation, Duke University, Durham

Teklehaimanot R, Wuhib E, Kassina A, Kidane Y, Alemu T, Spencer PS (1993) Patterns of Lathyrus sativus (grass pea) consumption and ODAP content of food samples in the lathyrism endemic regions of North West Ethiopia. Nutr Res 3:1113–1126

Urga K, Fite A, Kebede B (1995) Nutritional and anti-nutritional factors of grass pea germ plasm. Bull Chem Soc Ethiop 9:9–16

Urga K, Fufa H, Biratu E, Husain A (2005) Evaluation of Lathyrus sativus cultivated in Ethiopia for proximate composition, minerals, β-ODAP and anti-nutritional components. Afric J Food Agric Nutr Develop 5(1):1–15. Available https://www.ajol.info/index.php/ajfand/article/view/135979. Accessed 08th July 2020

Valamoti SM (2009) Plant food ingredients and 'recipes' from Prehistoric Greece: the archaeobotanical evidence. In: Morel JP, Mercuri AM (eds) Plants and culture: seeds of the cultural heritage of Europe. Bari, Edipuglia S.r.l., Bari, pp 25–38

Webb RE, Hyatt SA (1988) Haitian street foods and their nutritional contribution to dietary intake. Ecol Food Nutr 21(3):199–209. https://doi.org/10.1080/03670244.1988.9991033

Wilkins J, Hill S (2009) Food in the ancient world. Blackwell Publishing Ltd, Maiden, Oxford, and Carlton

Wilkinson J (2004) The food processing industry, globalization and developing countries. Electron J Agric Develop Econ 1(2):184–201

Chapter 3
Lathyrus sativus Cultivars and Grass Pea-Based Foods in Italy

Abstract The use of grain legumes in various European regions has been observed with peculiar interest in the Middle Age, although the presence of seeds such as *Lathyrus sativus* (grass pea) is demonstrated since the Neolithic Era. The cereal/legume association has been reported in Andorra, Italy, and France at least from the late Roman Empire to the late Middle Age. Cereals and other plants such as *Pisum sativum, Vicia faba,* and *L. sativus* were often present and probably cultivated in good amounts. Interestingly, the same localisation of known recipes based on *L. sativus* in certain areas found in Morocco, Spain, France, and Italy (as an ideal line from the Maghreb area to Central Europe) can be seen as the heritage of past memories. Anyway, grass pea is a known and cultivated grain legume in Europe and specifically in Italy. The presence and persistence of grass pea is demonstrated also because of the current existence of certain recipes in some areas with strong medieval traditions. This chapter concerns regional culinary traditions associated with grass pea in Italy, with some consideration concerning historical reasons for the localisation of *polenta*-like dishes in some circumscribed areas only.

Keywords Cicerchia · Heritage · *Lathyrus sativus* · Piciocia · Pisello breton · Polenta · Ru muocche

Abbreviations

EVOO	Extra-virgin olive oil
AECOSAN	The Spanish Agency for Consumer Affairs, Food Safety and Nutri-tion

© The Author(s), under exclusive license to Springer
Nature Switzerland AG 2020
M. Barone and R. Tulumello, *Lathyrus sativus and Nutrition,*
Chemistry of Foods, https://doi.org/10.1007/978-3-030-59091-8_3

3.1 Grass Pea in Italy. Diffusion

The use of grain legumes in various European regions has been observed with peculiar interest in the Middle Age, although the presence of seeds such as *Lathyrus sativus* (grass pea) is demonstrated since the Neolithic Era. The cereal/legume association (with relation to agricultural practices) has been reported in Andorra, Italy, and France at least (Bandini Mazzanti et al. 2005; Buonincontri et al. 2017; Cherubini 2000; Costantini et al. 1983; Esclassan et al. 2009; Montanari 2001) from the late Roman Empire to the late Middle Age. Cereals and other plants such as *Pisum sativum, Vicia faba, and Lathyrus sativus* were often present and probably cultivated in good amounts. Interestingly, the same localisation of known recipes based on *L. sativus* in certain areas found in Morocco, Spain, France, and Italy[1] can be seen as the heritage of past memories and the answer to old famines (Alfiero et al. 2017; Bell et al. 2014; Booth and Coveney 2007; Campbell 1997; Chong and Eun 1992; Cortese et al. 2016; de Suremain 2016; Delgado et al. 2017; Heinzelmann 2014; Helou 2006; Kollnig 2020; Larcher and Camerer 2015; Manning 2009; Maranzano 2014; McHugh 2015; Morton 2014; Patel et al. 2014; Pilcher 2017; Sheen 2010; Simopoulus and Bhat 2000; Steven 2018; Webb and Hyatt 1988; Wilkins and Hill 2009; Wilkinson 2004).

Anyway, grass pea is a known and cultivated grain legume in Europe and specifically in Italy (Fig. 3.1). The presence and persistence of grass pea is demonstrated also because of the current existence of certain recipes in some areas with strong medieval traditions. After all, one of the various names for grass pea in Italy is '*pisello bretonne*', probably meaning some relationship with ancient French civilisations in the past of Italian towns, especially in the Southern regions. Other food recipes have 'survived' until the present age, such as some traditional Sicilian or Genoese products made with (or containing) cereals or grain legumes (Abulafia 1978; Barbagli and Barzini 2010; Barone and Pellerito 2020; Carrabino 2017; CNC 2014a, b, c, d, e; Dauverd 2006; Guigoni 2004; Pisano 2011).

3.2 Grass Pea Cultivars and Soups in Italy

With relation to Italian traditions only, it has to be recognised that the use of grass pea is limited enough: in general, the most know recipes by the historical viewpoint at least are soups or similar dishes, and gruels or similar food preparations (Chap. 2).

The first category includes certainly a typical dish of Apulia: a soup of pulses and whole wheat. In this dish, the association between cereals (ready energy) and pulses (additional nutritional content related to protein at least) is evident. Used pulses are *cicerchia* (*L. sativus*), beans (*Phaseulus vulgaris*) lentils (*Lens culinaris*), peas

---

[1] As an ideal line from the Maghreb area to Central Europe.

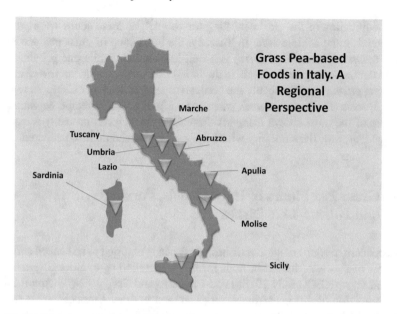

Fig. 3.1 With relation to Italian traditions, the use of grass pea is limited enough: in general, the most know recipes by the historical viewpoint at least are soups or similar dishes, and gruels or similar food preparations. Grass pea cultivars have also different typologies in Italy, being cited in the following regions at least: Lazio, Tuscany, Umbria, Marche, Abruzzo, Molise, Sicily, Sardinia, and Apulia. Interestingly, the use of *L. sativus* is highly mentioned in central Italian regions. With concern to Sicily, the use of local *cicerchia* legume is mentioned only in the medieval city of Troina, which was the first Sicilian capital at the early stages of the Norman County of Sicily

(*Pisum sativum*), favabeans (*Vicia faba*), and 'black garbanzo' beans. Interestingly, all of these grain legumes are typical cultivars of the local area, and grass pea cultivars have also different typologies in Italy, being cited in the following regions at least: Lazio, Tuscany, Umbria, Marche, Abruzzo, Molise, Sicily, Sardinia, and Apulia. One of the most known types, used here as a simple example because cultivars are a notable number, is the *cicerchia di Serra de'Conti* (Ancona, Italy). The interested Reader is invited to consult more specific literature in this ambit. The above-mentioned cultivar is considered here because of the use of this grain legume for the preparation of *vellutata di cicerchie* food: the main ingredient is naturally grass pea, with the addition of onions, carrots, milk cream, small slices of hard bread, salt and pepper, and typical parmesan or *pecorino* cheese. This recipe of Marche and Abruzzo regions demonstrates very well the possibility of soups and 'soup-like' dishes.

The 'soup' dish is present in Apulia, Molise, Umbria, and Marche regions at least with slightly different features: in general, differences concern the use of extra virgin olive oil (EVOO), garlic, onions, tomatoes, rosemary, other herbs, salt, pepper, potatoes, and also (in the Marche region) bacon or lard. This means that the nature of soups includes different ingredients, with the main presence of cereals and legumes, while other vegetable and animal ingredients may be allowed. Anyway, these are only new modification of the old 'meal for poor people', as these soups were probably considered in the Middle Age. In Tuscany, the local soup of grass pea and hulled wheat (*zuppa di cicerchia e farro*) is a well-known dish, while one peculiar *pasta* type in the Lazio Region (in detail, in the *Bassa Ciociaria* area) is '*laine e cicerchie*'. This long *pasta* type is probably the heritage of ancient Roman cuisine, as reported in the *Apicius' De Re Coquinaria* when speaking of a similar recipe: *laganum* (the ancestor of the most known *lasagna*). Actually, this recipe is probably a medieval heritage. Anyway, the similarity with *laine e cicerchie* should be considered.

3.3 Grass Pea Flours in Italy. Gruels, Porridges… and *Polenta*-Like Dishes

With concern to soups or similar dishes, grass pea is reported to be soaked and eaten in forms of stews, or turned in flour for the preparation of traditional (*gachas*) in Spain at least (AECOSAN 2018; Peña-Chocarro and Peña 1999; Valamoti 2009). The production of different gruels, very common in other European areas such as Central and Northern Italy (when *Cicer arietinum* flours are used for chick pea flat breads), demonstrates the large use of grass pea flours since the High Middle Age when speaking of breads from different flours and popular (and 'poor') soups (Cherubini 2000; Città di Troina 2017; Costantini et al. 1983; Montanari 2001).

With the exclusion of sauces—probably, an African tradition without similar traces in Europe—the use of grass pea flours is limited to the preparation of gruels porridges and other recipes such as Molise's '*Ru muocche of Poggio Sannita*'. This dish is very similar to the Northern Italian recipe of *polenta* or cornmeal mush: differently from common *polenta* (a yellow food obtained from maize flour), characteristic ingredients include grass pea, garlic, and chilli pepper.

The *vellutata di cicerchie* food is a soup-like dish with some similarities if compared with gruels and similar dishes. An interesting variety, at least from the historical viewpoint, can be found in Sicily, and in the Troina City in detail. In fact, the traditional soup with cereals and grain legumes was a typical dish of the Ancient Roman Empire, and the Sicilian province—historically known as the Roma's grain warehouse—was not an exception. In this region, a peculiar dish—the so-called *frascatula* of Sperlinga—corresponds to a creamy *polenta*-like dish (a possible missing ring between soups and polenta foods) based on wheat flour with addition of flours of various grain legumes, and several vegetables. Interestingly, the *frascatula* word derives from the Latin verbs *frascare* or *frangere,* and from the French word *flasque* (soft), while *polenta* may derive also from Latin word *pulus* or legume (Anonymous 2020; Sonnante et al. 2009; Soraci 2018). The recipe is typical of the inner areas of central Sicily; Sperlinga is in the Enna Province, the central area in Sicily. Different versions of the *frascatula* food are present in the same area: the *paniccia* recipe (Enna), the *picciotta* type in Modica, and other typologies may be found in other provinces of the same Region. Interestingly, a particular variety in the Troina area (province of Enna) is the *piciocia*, and the peculiar feature is the presence of *L. sativus* as the main legume (Anonymous 2003–2012–2020a, b). Substantially, this *polenta*-like dish is prepared using various legume flours (chickpeas, peas, grass pea), where *L. sativus* is the main legume, 54% of the total amount of flour). Wild cardoons (*Cynara cardunculus* var. *sylvestris* Lam.), bacon, onions, and chilli pepper can be used (Anonymous 2003–2020a, b).

A final historical reflection should concern specifically *piciocia*. In fact, the main and characteristic ingredient is grass pea, also named (in Italy) *pisello bretonne* (Chap. 1). The exact translation from Italian language is 'pea of Bretagne', while 'Bretagne' is naturally a historical region of Northern France. Interestingly, it is known that:

(a) The Kingdom of Bretagne has been conquered by the Normans in 913 B.C.
(b) The Arab Emirate of Sicily has been also conquered between 1061 and 1091 B.C. by the Normans.

The common point between these two important facts is the presence of Normans before in Bretagne, and after in Sicily (more than a century passed between the two invasions). At the same time, it should be remembered that the presence of grass pea in Sicily and specifically in Troina at least is not an exception, while the preparation of a *polenta*-like dish with grass pea (the 'pea of Bretagne') is the only example in central Sicily at least. Could this recipe be one of heritages of the Norman domination in Sicily? This hypothesis has to be naturally evaluated carefully from a historical viewpoint, and there are no peculiar evidences in our knowledge so far… However, it is curious that the only example of *polenta*-like dish with *L. sativus* is present only in Troina, the first capital city of the new Norman Countship of Sicily (1062 B.C.). This reflection might confirm the historical and medieval tradition of many grass pea-based recipes in the medieval Europe until now.

References

Abulafia D (1978) Pisan commercial colonies and consulates in twelfth-century Sicily. Eng Historical Rev 93(366):68–81

AECOSAN (2018) Report of the Scientific Committee of the Spanish Agency for Con-sumer Affairs, Food Safety and Nutrition (AECOSAN) on the safety of grass pea flour consumption. Revista del Comité Científico de la AECOSAN 27: 61–78. The Spanish Agency for Consumer Affairs, Food Safety and Nutrition (AECOSAN), Madrid

Alfiero S, Giudice AL, Bonadonna A (2017) Street food and innovation: the food truck phenomenon. Brit Food J 119(11):2462–2476. https://doi.org/10.1108/BFJ-03-2017-0179

Anonymous (2003) Piciocia—(Polenta troinese). Studio d'Ingegneria Massimiliano Stazzone (SIMS). Available https://www.stazzone.it/troina/cultura/gastronomia/cucina_tipica/piciocia.htm. Accessed 14th July 2020

Anonymous (2012) Piciocia siciliana. Hysteria Lane. Available https://cookingbreakdown.blogspot.com/2012/10/piciocia-siciliana.html. Accessed 14th July 2020

Anonymous (2013) Vellutata di cicerchia…. un sapore antico. L'ingrediente perduto. Available https://ingredienteperduto.blogspot.com/2013/01/vellutata-di-cicerchiaun-sapore-antico.html. Accessed 14th July 2020

Anonymous (2020a) Frascatula, polenta siciliana, storia e varianti. Giallozafferano. Available https://blog.giallozafferano.it/vittoriaaifornelli/frascatula-polenta. Accessed 14th July 2020

Anonymoys (2020b) Passatoinrete–F–Frascàtula. IT&T srl. Available https://www.passatoinretepr esente.it/f/. Accessed 14th July 2020

Bandini Mazzanti M, Bosi G, Mercuri AM, Accorsi CA, Guarnieri C (2005) Plant use in a city in Northern Italy during the late Mediaeval and Renaissance periods: results of the archaeobotanical investigation of "The Mirror Pit" (14th–15th century a.d.) in Ferrara. Veget Hist Archaeobot 14:442–452. https://doi.org/10.1007/s00334-005-0082-y

Barbagli A, Barzini S (2010) Pane, pizze e focacce. Giunti Editore, Florence

Bell JS, Loukaitou-Sideris A (2014) Sidewalk informality: an examination of street vending regulation in China. Int Plan Stud 19(3–4):221–243. https://doi.org/10.1080/13563475.2014. 880333

Booth SL, Coveney J (2007) Survival on the streets: prosocial and moral behaviors among food insecure homeless youth in Adelaide, South Australia. J Hunger Environ Nutr 2(1):41–53. https:// doi.org/10.1080/19320240802080874

Buonincontri MP, Pecci A, Di Pasquale G, Ricci P, Lubritto C (2017) Multiproxy approach to the study of Medieval food habits in Tuscany (central Italy). Archaeol Anthropol Sci 9(4):653–671. https://doi.org/10.1007/s12520-016-0428-7

Campbell CG (1997) Grass pea, *Lathyrus sativus* L. Promoting the conservation and use of under-utilized and neglected crops. Rome, Italy: Institute of Plant Genetics and Crop Plant Research, Rome, and Germany/International Plant Genetic Resources Institute, Gatersleben, Nr 18. p 92

Carrabino D (2017) 14 Oratories of the Compagnie of Palermo: sacred spaces of rivalry. In: Bullen Presciutti D (ed) Space, place, and motion: locating confraternities in the late Medieval and Early Modern City. Koninklijke Brill Nv (Brill), Leiden, pp 344–371. https://doi.org/10.1163/978900 4339521_016

Cherubini G (2000) L'approvvigionamento alimentare delle città toscane tra il XII e il XIV secolo. Rivista di storia dell'agricoltura 40:33–52

Chong HK, Eun VLN (1992) 4 Backlanes as contested regions: construction and control of physical. In: Huat CB, Edwards N (eds) Public space: design, use and management. Singapore University Press, Singapore

Città di Troina (2017) Sottoscritto protocollo d'intesa con il Consiglio Nazionale dei Chimici per la salvaguardia e la garanzia della tradizione gastronomica di Troina. Available https://www.com une.troina.en.it/DETTAGLIO_NEWS.ASP?ID=470. Accessed 14th July 2020

CNC (2014a) Accordo tra Comune di Palermo e Consiglio Nazionale dei Chimici (CNC) per la partecipazione alla realizzazione di un sistema di salvaguardia e garanzia della tradizione gastro-nomica palermitana', Prot. 646/14/cnc/fta. Consiglio Nazionale dei Chimici (CNC), Rome. Avail-able https://www.chimicifisici.it/wp-content/uploads/2018/10/20131210_accordo_firmato_dal_ Presidente_del_CNC.pdf. Accessed 07 April 2020

CNC (2014b) DOMANDA DI REGISTRAZIONE - Art. 8 - Regolamento (UE) n. 1151/2012 del Parlamento Europeo e del Consiglio del 21 novembre 2012 sui regimi di qualità dei prodotti agricoli e alimentari - " "ARANCINA". Annex to the document 'Accordo tra Comune di Palermo e Consiglio Nazionale dei Chimici (CNC) per la partecipazione alla realizzazione di un sistema di salvaguardia e garanzia della tradizione palermitana', Prot. 646/14/cnc/fta. Consiglio Nazionale dei Chimici (CNC), Rome. Available https://www.chimicifisici.it/wp-con tent/uploads/2018/10/ARANCINA_CNC__STG_2014.pdf. Accessed 08 Apr 2020

CNC (2014c) DOMANDA DI REGISTRAZIONE - Art. 8 - Regolamento (UE) n. 1151/2012 del Parlamento Europeo e del Consiglio del 21 novembre 2012 sui regimi di qualità dei prodotti agri-coli e alimentari - "SFINCIONELLO". Annex to the document 'Accordo tra Comune di Palermo e Consiglio Nazionale dei Chimici (CNC) per la partecipazione alla realizzazione di un sistema di salvaguardia e garanzia della tradizione gastronomica palermitana', Prot. 646/14/cnc/fta. Consiglio Nazionale dei Chimici (CNC), Rome. Available https://www.chimicifisici.it/wp-con tent/uploads/2018/10/SFINCIONELLO_CNC_STG_2014.pdf. Accessed 07 Apr 2020

CNC (2014d) DOMANDA DI REGISTRAZIONE - Art. 8 - Regolamento (UE) n. 1151/2012 del Parlamento Europeo e del Consiglio del 21 novembre 2012 sui regimi di qualità dei prodotti agricoli e alimentari - "PANE CA MEUSA". Annex to the document 'Accordo tra Comune di Palermo e Consiglio Nazionale dei Chimici (CNC) per la partecipazione alla realizzazione di un sistema di salvaguardia e garanzia della tradizione gastronomica palermitana', Prot. 646/14/cnc/fta. Consiglio Nazionale dei Chimici (CNC), Rome. Available https://www.chimicifisici.it/wp-con tent/uploads/2018/10/PANE_CA_MEUSA_rev.3dic2014.pdf. Accessed 08 Apr 2020

CNC (2014e) DOMANDA DI REGISTRAZIONE DI UNA STG - Art. 8 - Regolamento (UE) n. 1151/2012 del Parlamento Europeo e del Consiglio del 21 novembre 2012 sui regimi di qualità dei prodotti agricoli e alimentary - "PANE E PANELLE". Annex to the document 'Accordo tra Comune di Palermo e Consiglio Nazionale dei Chimici (CNC) per la partecipazione alla realizzazione di un sistema di salvaguardia e garanzia della tradizione gastronomica palermitana', Prot. 646/14/cnc/fta. Consiglio Nazionale dei Chimici (CNC), Rome.Available https://www.chimicifi sici.it/wp-content/uploads/2018/10/PANE_E_PANELLE__CNC_STG_2014.pdf. Accessed 07 Apr 2020

Cortese RDM, Veiros MB, Feldman C, Cavalli SB (2016) Food safety and hygiene practices of vendors during the chain of street food production in Florianopolis, Brazil: a cross-sectional study. Food Control 62:178–186. https://doi.org/10.1016/j.foodcont.2015.10.027

Costantini L, Costantini L, Napolitano G, Whitehouse D (1983) Cereali e legumi medievali provenienti dalle mura di Santo Stefano, Anguillara Sabazia (Roma). Archeologia Medievale 10:393–413

Dauverd C (2006) Genoese and Catalans: trade diaspora in Early Modern Sicily. Mediterr Stud 15:42–61

de Suremain CÉ (2016) The never-ending reinvention of 'traditional food'. In: Sébastia B (ed) Eating traditional food: politics, identity and practices. Routledge, Abingdon

Delgado AM, Almeida MDV, Parisi S (2017) Chemistry of the Mediterranean Diet. Springer International Publishing, Cham. https://doi.org/10.1007/978-3-319-29370-7

Esclassan R, Grimoud AM, Ruas MP, Donat R, Sevin A, Astie F, Lucas S, Crubezy E (2009) Dental caries, tooth wear and diet in an adult medieval (12th–14th century) population from Mediterranean France. Arch Oral Biol 54(3):287–297. https://doi.org/10.1016/j.archoralbio.2008.11.004

Guigoni A (2004) La cucina di strada Con una breve etnografia dello street food genovese. Mneme-Revista de Humanidades 4, 9 fev./mar. de 2004:32–43

Heinzelmann U (2014) Beyond bratwurst: a history of food in Germany. Reaktion Books Ltd., London

Helou A (2006) Mediterranean street food: stories, soups, snacks, sandwiches, barbecues, sweets, and more from Europe, North Africa, and the Middle East. Harper Collins Publishers, New York

Kollnig S (2020) The 'good people'of Cochabamba city: ethnicity and race in Bolivian middle-class food culture. Lat Am Caribb Ethn Stud 15(1):23–43. https://doi.org/10.1080/17442222.2020.169 1795

Lanari M (2019) Legumi da riscoprire: la cicerchia. La Cucina Italiana. https://www.lacucinaitalian a.it. Available https://www.lacucinaitaliana.it/news/trend/insetti-in-tavola-un-italiano-su-due-li-assaggerebbe/. Accessed 14th July 2020

Larcher C, Camerer S (2015) Street Food. Temes de Dissen 31:70–83. Available https://core.ac.uk/download/pdf/39016176.pdf. Accessed 09 Apr 2020

Manning JA (2009) Constantly containing. Dissertation, West Virginia University

Maranzano B (2014) Lo sviluppo del fenomeno "street food": il cibo di strada a Palermo ieri e oggi. Dissertation, University of Pisa, Italy

McHugh MR (2015) Modern Palermitan markets and street food in the Ancient Roman World. Conference paper, the Oxford Symposium on Food and Cookery, St. Catherine's College, Oxford University

Montanari M (2001) Cucina povera, cucina ricca. Quaderni Medievali 52:95–105

Morton PE (2014) Tortillas: a cultural history. University of New Mexico Press, Albuquerque

Patel K, Guenther D, Wiebe K, Seburn RA (2014) Promoting food security and livelihoods for urban poor through the informal sector: a case study of street food vendors in Madurai, Tamil Nadu, India. Food Sec 6(6):861–878. https://doi.org/10.1007/s12571-014-0391-z

Peña-Chocarro L, Peña LZ (1999) History and traditional cultivation of *Lathyrus sativus* L. and *Lathyrus citera* L. in the Iberian peninsula. Veget Hist Archaeobot 8(1–2):49–52. https://doi.org/10.1007/BF02042842

Pilcher JM (2017) Planet taco: a global history of Mexican food. Oxford University Press, Oxford

Pisano A (2011) La farinata diventa fainé. Un esempio di indigenizzazione. Intrecci. Quaderni di antropologia cultural I 1:35–58. Associazione Culturale Demo Etno Antropologica, Sassari

Sheen B (2010) Foods of Egypt. Greenhaven Publishing LLC, New York

Simopoulus AP, Bhat RV (2000) Street foods. Karger AG, Basel

Sonnante G, Hammer K, Pignone D (2009) From the cradle of agriculture a handful of lentils: history of domestication. Rend Fis Acc Lincei 20(1):21–37. https://doi.org/10.1007%2Fs12210-009-0002-7

Soraci L (2018) Frascatura siciliana ovvero la polenta siciliana. Profumo di Sicilia, https://www.profumodisicilia.net. Available https://www.profumodisicilia.net/2018/01/02/frascatura-siciliana/. Accessed 14th July 2020

Steven QA (2018) Fast food, street food: Western fast food's influence on fast service food in China. Dissertation, Duke University, Durham

Valamoti SM (2009) Plant food ingredients and 'recipes' from Prehistoric Greece: the archaeobotanical evidence. In: Morel JP, Mercuri AM (eds) Plants and culture: seeds of the cultural heritage of Europe. Bari, Edipuglia S.r.l., Bari, pp 25–38

Webb RE, Hyatt SA (1988) Haitian street foods and their nutritional contribution to dietary intake. Ecol Food Nutr 21(3):199–209. https://doi.org//10.1080/03670244.1988.9991033

Wilkins J, Hill S (2009) Food in the ancient world. Blackwell Publishing Ltd, Maiden, Oxford, and Carlton

Wilkinson J (2004) The food processing industry, globalization and developing countries. Electron J Agric Dev Econ 1(2):184–201

Chapter 4
Grass Pea, the β-ODAP Toxin, and Neurolathyrism. Health and Safety Considerations

Abstract This Chapter concerns the use of grass pea (*Lathyrus sativus*) in human and animal nutrition by the viewpoint of safety and health. Domesticated grass pea is documented and historically demonstrated. This grain legume can be found worldwide. Substantially, *Lathyrus* genus—and grass pea in particular—seems to be a constant presence in the human history in association with cultivated cereals and other legumes. From the nutritional viewpoint, *L. sativus* is reported to have a demonstrated deficiency of vitamins A, B_1, C, D, and E, some amino acids and elements. However, these deficiencies are not important worries. Health and safety problems are linked to the β-ODAP neurotoxin, responsible for neurolathyrism, and other antinutritional factors (condensed tannins and total phenolics; some enzymatic inhibitory activities). Actually, 'neurolathyrism is associated with osteolathyrism (under the common 'lathyrism' name). Anyway, the main problem is the presence of β-ODAP neurotoxin, and a safe use of grass pea cannot be assured in these conditions. This Chapter also explains common culinary methods for the safe consumption of grass pea-based foods by the chemical angle.

Keywords α-ODAP · β-ODAP · BAPN · Grass pea · *Lathyrus sativus* · Neurolathyrism · Osteolathyrism

Abbreviations

ODAP   Diaminopropionic acid
BAPN   β-Aminopropionitrile toxin
BOAA   β-*N*-oxalyl-ammo-L-alanine
β-ODAP β-*N*-Oxalyl-L-α-β-diaminopropionic acid

© The Author(s), under exclusive license to Springer
Nature Switzerland AG 2020
M. Barone and R. Tulumello, *Lathyrus sativus and Nutrition*,
Chemistry of Foods, https://doi.org/10.1007/978-3-030-59091-8_4

4.1   Lathyrus Sativus and Neurolathyrism. Health and Safety Concerns

The use of grass pea (*Lathyrus sativus*) in human and animal nutrition is documented and historically demonstrated because of its capability to survive in extremely difficult environments. This grain legume can be found in all continents: North and the Sub-Saharan area (Africa), India, Bangladesh, Nepal Pakistan, and the Middle East (Asia), Southern Europe, the Americas, and—more recently—Australia.

Substantially, *Lathyrus* genus—and grass pea in particular—seems to be a constant presence in the human history in association with cultivated cereals and other legumes. Historically, the medieval cuisine in Europe has many traces of grass pea, and certain culinary traditions survive in various forms (Alfiero et al. 2017; Bell and Loukaitou-Sideris 2014; Booth and Coveney 2007; Chong and Eun 1992; Cortese et al. 2016; de Suremain 2016; Delgado et al. 2017; Heinzelmann 2014; Helou 2006; Kollnig 2020; Larcher and Camerer 2015; Manning 2009; Maranzano 2014; McHugh 2015; Morton 2014; Patel et al. 2014; Pilcher 2017; Sheen 2010; Simopoulus and Bhat 2000; Steven 2018; Webb and Hyatt 1988; Wilkins and Hill 2009; Wilkinson 2004). The cereals/grain legumes association is also recognised in the old Europe of Middle Age (Abulafia 1978; Barbagli and Barzini 2010; Carrabino 2017; CNC 2014a, b, c, d, e; Dauverd 2006; Guigoni 2004; Pisano 2011).

L. sativus is reported to have a demonstrated deficiency of vitamins A, B_1, C, D, and E, some amino acids (isoleucine, lysine, methionine, sodium chloride, fat matter, and some elements: cobalt, phosphorus, sulphur, fluorine, and iodine (Cocks et al. 2000; Enneking 2011; López Aydillo and Toledano Jiménez Castellanos 1968). However, these deficiencies are not important worries: grass pea is also reported to be pleasant from the organoleptic viewpoint, and it is also rich of protein (26–30%) and iron. Other antinutritional factors may be the profile of condensed tannins and total phenolics which might cause astringency and associated bitterness, and some enzymatic inhibitory activities with peculiar reference to trypsin and chymotrypsin.

The 'bad' reputation of *L. sativus* is associated with a peculiar disease, 'neurolathyrism', even if this illness may be also found together with another disease, the *konzo* illness ('tied legs'), caused by ingestion of roots of *Manihot esculenta* (cassava) roots (Gresta et al. 2014; Lambein et al. 2019; Nzwalo and Cliff 2011).

Actually, the 'lathyrism' disease is associated with the whole *Lathyrus* genus. In general, the term implies permanent paralysis of the lower limbs in humans and in animals, specifically ruminants and monogastric species. The cause is a water-soluble neurotoxin: β-*N*-oxalyl-L-α-β-diaminopropionic acid (β-ODAP, on condition that it is assumed for a prolonged time (Campbell et al. 1994; Cocks et al. 2000).

Actually, 'lathyrism' may be referred to two different paralysis diseases: osteolathyrism and neurolathyrism. The first phenomenon has one specific target: bones (skeletal deformities), and it is generally reported to be caused by the consumption of four *Lathyrus* species: *L. odoratus,L. hirsutus,L. pusillus,* and *L. roseus.* In this ambit, it has been reported that the main cause is β-aminopropionitrile toxin (BAPN), found also in *L. sativus* and *L. cicera* provided that interested subjects suffer also from chronic neurolathyrism. However, BAPN is not the main target of grass pea-associated concerns when speaking of health and safety.

4.2 β-ODAP. Chemistry, Associated Concerns and Possible Solutions

The main responsible for neurolathyrism, β-ODAP, acts as a glutamate analogue when speaking of reactions of the nervous system. The neurotoxin is probably able to create strong bonds with several glutamate receptors with consequent permanent damages because of neuron degenerations (Hanbury et al. 2000; Hugon et al. 2000). Interestingly, effects on the animal system are not important enough if compared with damages in humans, probably because of zinc deficiency (similarly to plants), and with the delayed growth in some yeasts and insects (so, β-ODAP appears to be an antifeedant substance) (Lambein and Kuo 1997).

By the chemical viewpoint, β-ODAP—also named β-N-oxalyl-ammo-L-alanine (BOAA) or dencichin, molecular weight: 176, molecular formula: $C_5H_8N_2O_5$, Fig. 4.1 (Campbell et al. 1994; Lambein et al. 2019; Xie et al. 2007)—is obtained in *L. sativus* from the heterocyclic amino acid β-isoxazolin-5-on-2-yl-alanine. This enzymatic synthesis occurs with a remarkable amount only when seeds are produced in *Pisum sativum* and *Lens culinaris* (Kuo et al. 2003; Lambein et al. 2019).

It has also been reported that β-ODAP is synthesised when one or more of the following factors are observed: drought, lack of zinc, and excessive iron amount (in soils). Also, the enzymatic conversion turning β-isoxazolin-5-on-2-yl-alanine into β-ODAP is not clear at present (Lambein et al. 2019). With reference to analytical detection methods, and considering that not only *L. sativus* but also non-legume vegetables such as *Panax notoginseng* (Burk.) can contain β-ODAP, some techniques have been proposed so far, including a colorimetric system (reagents: o-phthalaldehyde with

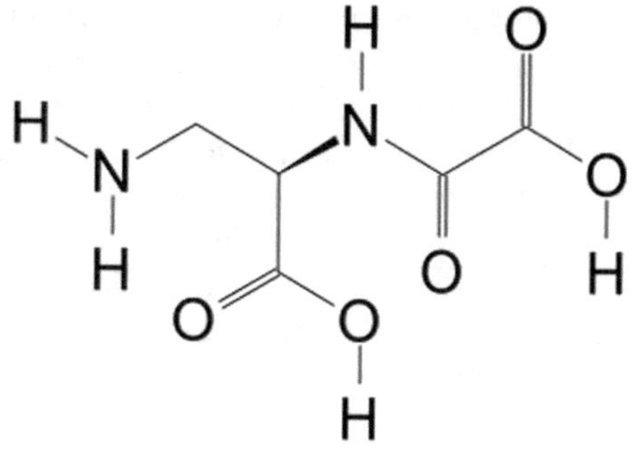

Fig. 4.1 α-ODAP form, in comparison with β-ODAP (Fig. 1.2)

α,β-diaminopropionic acid) and different chromatographic methods (Liu et al. 2017; Xie et al. 2007).

It has to be considered that ODAP is present in two forms: Fig. 4.1 shows also the α-isomer which is believed to be less toxic than β-ODAP. Consequently, toxicity depends mainly on the amount of the β-isomer on the total diaminopropionic (ODAP) quantity. Substantially, the β-isomer is reported to slowly reach the equilibrium with the α-form, more easily heating; in nature, the β:α ratio is approximately 95:5, and the ratio can be modified in different conditions with different results. Anyway, the α-isomer does not seem to give neurotoxic reactivity during in vitro tissue tests (Yan et al. 2006).

Possible strategies against lathyrism are generally based on heating treatments (Chaps. 1, 3, and 4), in connection with soaking technique (β-ODAP is soluble in water, and its content may be lowered by means of simple aqueous 'washing' and consequent extraction). It has been demonstrated that autoclaving on feed containing β-ODAP (Fig. 4.1) can really lower the amount of this toxin until 30–90% (Akalu et al. 1998; Hanbury et al. 2000; Teklehaimanot et al. 1993), but more research is needed. In fact, the real effect of boiling treatments is not clear enough at present because the toxin is water-soluble. Consequently, the removal of β-ODAP might be also caused by simple aqueous extraction. On the other hand, roasting is reported to be really effective in this ambit because of the possible isomerisation of β-ODAP in the α-ODAP form (which is not active, Fig. 4.1) with a possible 60:40 ratio between β- and α-isomers (Akalu et al. 1998; Hanbury et al. 2000). Because of the high variability in neurotoxin amounts for the same grass pea cultivars worldwide (Cocks et al. 2000; Hanbury et al. 1999), it is hard to establish a real analytical limit for lathyrism dangers. It is known that high β-ODAP amounts are produced as the consequence of 'water stress', and that the increase of the toxin in seeds is apparently correlated with the diminution of total β-ODAP in the whole plant. Substantially, the easier the growth (of the vegetable under non-stressed conditions, with augment of seeds), the lower the toxin production. Anyway, the correlation between low β-ODAP production and higher sowing dates and water stress conditions has been confirmed, but the role of genetic features is not clear enough at present (Cocks et al. 2000).

Some food processing (basically: soaking, heating, and soaking + cooking) procedures have been studied in different papers (Akalu et al. 1998; Jha 1987; Spencer and Palmer 2003; Teklehaimanot et al. 1993, 2005) with the aim of obtaining a general β-ODAP reduction from 30 to 88% (Fig. 4.2).

Because of the risk of 50% and more of the neurotoxic form in the total ODAP amount, it is probable that the best strategy is the production of less active L. sativus species by means of genetic and microbiological studies, and/or improved agricultural practices. With concern to the first field, it has to be considered that rumen microflora appears able to destroy β-ODAP, and this strategy may be useful if used sinergically with other options. Also, the use of grass pea with low β-ODAP amounts may be useful when speaking of the production of animal foods for human consumption (because the neurotoxic 'answer' of humans is more worrying than in animals).

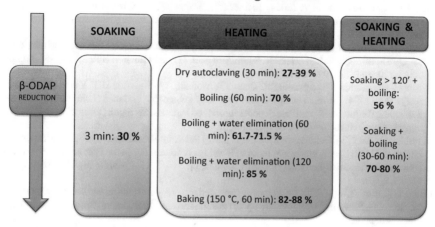

Fig. 4.2 Some food processing (basically: soaking, heating, and soaking + cooking) procedures have been studied in different papers (Akalu et al. 1998; Jha 1987; Spencer and Palmer 2003; Teklehaimanot et al. 1993, 2005) with the aim of obtaining a general β-ODAP reduction from 30 to 88%. It has to be considered that experiments have been carried out on husked or whole seeds, or on grass pea flours. Consequently, the nature of examined samples can determine different results when speaking of the same technique

Genetic studies have also allowed the improvement of new grain legumes belonging to *L. sativus* with enhanced resistance to biotic and abiotic menaces, and with consequent reduced amount of produced β-ODAP in seeds (because seeds are the part with the most important amount). Techniques such as intra-specific hybridisation have allowed good results, and the production of varieties with <0.1% β-ODAP < 0.1% is now possible. However, the production of low-toxic grass pea foods remains currently an open problem even if the characterisation of germplasm collections by means of molecular markers is a promising road (Kumar et al. 2013). As a result, grass pea has good perspectives with relation to possible uses for human and animal nutrition, but more studies are surely needed at present worldwide.

References

Abulafia D (1978) Pisan commercial colonies and consulates in twelfth-century Sicily. Eng Hist Rev 93(366):68–81

Akalu G, Johansson G, Nair BM (1998) Effect of processing on the content of β-N-oxalyl-α, β-diaminopropionic acid (gb-ODAP) in grass pea (*Lathyrus sativus*) seeds and flour as determined by flow injection analysis. Food Chem 62(2):233–237. https://doi.org/10.1016/S0308-8146(97)00137-4

Alfiero S, Giudice AL, Bonadonna A (2017) Street food and innovation: the food truck phenomenon. Brit Food J 119(11):2462–2476. https://doi.org/10.1108/BFJ-03-2017-0179

Barbagli A, Barzini S (2010) Pane, pizze e focacce. Giunti Editore, Florence

Bell JS, Loukaitou-Sideris A (2014) Sidewalk informality: An examination of street vending regulation in China. Int Plan Stud 19(3–4):221–243. https://doi.org/10.1080/13563475.2014. 880333

Booth SL, Coveney J (2007) Survival on the streets: prosocial and moral behaviors among food insecure homeless youth in Adelaide, South Australia. J Hunger Environ Nutr 2(1):41–53. https:// doi.org/10.1080/19320240802080874

Campbell CG, Mehra RB, Agrawal SK, Chen YZ, Moneim AA, Khawaja HIT, Yadov CR, Tay JU, Araya WA (1994) Current status and future strategy in breeding grasspea (Lathyrus sativus). In: Expanding the production and use of cool season food legumes. Springer, Dordrecht, pp 617–630. https://doi.org/10.1007/978-94-011-0798-3_37

Carrabino D (2017) 14 Oratories of the Compagnie of Palermo: Sacred Spaces of Rivalry. In: Bullen Presciutti D (ed) Space, place, and motion: locating confraternities in the late medieval and early modern city. Koninklijke Brill Nv (Brill), Leiden, pp 344–371. https://doi.org/10.1163/978900 4339521_016

Chong HK, Eun VLN (1992) 4 Backlanes as contested regions: construction and control of physical. In: Huat CB, Edwards N (eds) Public space: design, use and management. Singapore University Press, Singapore

CNC (2014a) Accordo tra Comune di Palermo e Consiglio Nazionale dei Chimici (CNC) per la partecipazione alla realizzazione di un sistema di salvaguardia e garanzia della tradizione gastronomica palermitana', Prot. 646/14/cnc/fta. Consiglio Nazionale dei Chimici (CNC), Rome. Available https://www.chimicifisici.it/wp-content/uploads/2018/10/20131210_accordo_firmato_dal_Presidente_del_CNC.pdf. Accessed 07 Apr 2020

CNC (2014b) DOMANDA DI REGISTRAZIONE—Art. 8—Regolamento (UE) n. 1151/2012 del Parlamento Europeo e del Consiglio del 21 novembre 2012 sui regimi di qualità dei prodotti agricoli e alimentari—"ARANCINA". Annex to the document 'Accordo tra Comune di Palermo e Consiglio Nazionale dei Chimici (CNC) per la partecipazione alla realizzazione di un sistema di salvaguardia e garanzia della tradizione gastronomica palermitana', Prot. 646/14/cnc/fta. Consiglio Nazionale dei Chimici (CNC), Rome. Available https://www.chimicifisici.it/wp-con tent/uploads/2018/10/ARANCINA_CNC__STG_2014.pdf. Accessed 08 Apr 2020

CNC (2014c) DOMANDA DI REGISTRAZIONE—Art. 8—Regolamento (UE) n. 1151/2012 del Parlamento Europeo e del Consiglio del 21 novembre 2012 sui regimi di qualità dei prodotti agricoli e alimentari—"SFINCIONELLO". Annex to the document 'Accordo tra Comune di Palermo e Consiglio Nazionale dei Chimici (CNC) per la partecipazione alla realizzazione di un sistema di salvaguardia e garanzia della tradizione gastronomica palermitana', Prot. 646/14/cnc/fta. Consiglio Nazionale dei Chimici (CNC), Rome. Available https://www.chimicifisici.it/wp-con tent/uploads/2018/10/SFINCIONELLO_CNC_STG_2014.pdf. Accessed 07 Apr 2020

CNC (2014d) DOMANDA DI REGISTRAZIONE—Art. 8—Regolamento (UE) n. 1151/2012 del Parlamento Europeo e del Consiglio del 21 novembre 2012 sui regimi di qualità dei prodotti agricoli e alimentari—"PANE CA MEUSA". Annex to the document 'Accordo tra Comune di Palermo e Consiglio Nazionale dei Chimici (CNC) per la partecipazione alla realizzazione di un sistema di salvaguardia e garanzia della tradizione gastronomica palermitana', Prot. 646/14/cnc/fta. Consiglio Nazionale dei Chimici (CNC), Rome. Available https://www.chimicifisici.it/wp-con tent/uploads/2018/10/PANE_CA_MEUSA_rev.3dic2014.pdf. Accessed 08 Apr 2020

CNC (2014e) DOMANDA DI REGISTRAZIONE DI UNA STG—Art. 8—Regolamento (UE) n. 1151/2012 del Parlamento Europeo e del Consiglio del 21 novembre 2012 sui regimi di qualità dei prodotti agricoli e alimentary—"PANE E PANELLE". Annex to the document 'Accordo tra Comune di Palermo e Consiglio Nazionale dei Chimici (CNC) per la partecipazione alla realizzazione di un sistema di salvaguardia e garanzia della tradizione gastronomica palermitana', Prot. 646/14/cnc/fta. Consiglio Nazionale dei Chimici (CNC), Rome. Available https://www.chimicifi

sici.it/wp-content/uploads/2018/10/PANE_E_PANELLE__CNC_STG_2014.pdf. Accessed 07 Apr 2020

Cocks P, Siddique K, Hanbury C (2000) Lathyrus: a new legume. A Report for the Rural Industries Research and Development Corporation, Jan 2000, RIRDC Publication No 99/150, RIRDC Project No UWA-21A. Rural Industries Research & Development Corporation (RIRDC), now AgriFutures Australia, Wagga Wagga

Cortese RDM, Veiros MB, Feldman C, Cavalli SB (2016) Food safety and hygiene practices of vendors during the chain of street food production in Florianopolis, Brazil: a cross-sectional study. Food Control 62:178–186. https://doi.org/10.1016/j.foodcont.2015.10.027

Dauverd C (2006) Genoese and catalans: trade diaspora in early modern Sicily. Mediterr Stud 15:42–61

de Suremain CÉ (2016) The never-ending reinvention of 'traditional food'. In: Sébastia B (ed) Eating traditional food: politics, identity and practices. Routledge, Abingdon

Delgado AM, Almeida MDV, Parisi S (2017) Chemistry of the mediterranean diet. Springer International Publishing, Cham. https://doi.org/10.1007/978-3-319-29370-7

Enneking D (2011) The nutritive value of grasspea (*Lathyrus sativus*) and allied species, their toxicity to animals and the role of malnutrition in neurolathyrism. Food Chem Toxicol 49(3):694–709. https://doi.org/10.1016/j.fct.2010.11.029

Gresta F, Rocco C, Lombardo GM, Avola G, Ruberto G (2014) Agronomic characterization and α-and β-ODAP determination through the adoption of new analytical strategies (HPLC-ELSD and NMR) of ten Sicilian accessions of grass pea. J Agric Food Chem 62(11):2436–2442. https://doi.org/10.1021/jf500149n

Guigoni A (2004) La cucina di strada Con una breve etnografia dello street food genovese. Mneme-Revista de Humanidades 4, 9 fev./mar. de 2004:32–43

Hanbury CD, Siddique KHM, Galwey NW, Cocks PS (1999) Genotype-environment interaction for seed yield and ODAP concentration of *Lathyrus sativus* L. and *L. cicera* L. in Mediterranean-type environments. Euphytica 110(1):45–60. https://doi.org/10.1023/A:1003770216955

Hanbury CD, White CL, Mullan BP, Siddique KHM (2000) A review of the potential of *Lathyrus sativus* L. and *L. cicera* L. grain for use as animal feed. Anim Feed Sci Technol 87:1–27. https://doi.org/10.1016/S0377-8401(00)00186-3

Heinzelmann U (2014) Beyond bratwurst: a history of food in Germany. Reaktion Books Ltd., London

Helou A (2006) Mediterranean street food: stories, soups, snacks, sandwiches, barbecues, sweets, and more from Europe, North Africa, and the Middle East. Harper Collins Publishers, New York

Hugon J, Ludolph AC, Spencer PS (2000) β-N-oxalylamino-alanine. In: Spencer PS, Schaumburg H (eds) Experimental and clinical Neurotoxicology, 2nd edn. Oxford University Press, New York, pp 925–938

Jha K (1987) Effect of the boiling and decanting method of Khesari (*Lathyrus sativus*) detoxification, on changes in selected nutrients. Arch Latinoam Nutr 37:101–107

Kollnig S (2020) The 'good people' of Cochabamba city: ethnicity and race in Bolivian middle-class food culture. Lat Am Caribb Ethn Stud 15(1):23–43. https://doi.org/10.1080/17442222.2020.1691795

Kumar S, Gupta P, Barpete S, Sarker A, Amri A, Mathur PN, Baum M (2013) Grass pea. In: Singh M, Upadhyaya HD, Singh Bisht (eds) Genetic and Genomic Resources of Grain Legume Improvement. Elsevier, London and Walthan, pp 269–292. https://doi.org/10.1016/B978-0-12-397935-3.00011-6

Kuo YH, Ikegami F, Lambein F (2003) Neuroactive and other free amino acids in seed and young plants of Panax ginseng. Phytochem 62:1087–1091. https://doi.org/10.1016/S0031-9422(02)00658-1

Lambein F, Kuo YH (1997) Lathyrus sativus, a neolithic crop with a modern future? an overview of the present situation. In: Proceedings of the international conference '*Lathyrus sativus*— cultivation and nutritional value in animals and human', Lublin, Radom, Poland, 9–10 June 1997, pp 6–12

Lambein F, Travella S, Kuo YH, Van Montagu M, Heijde M (2019) Grass pea (*Lathyrus sativus* L.): orphan crop, nutraceutical or just plain food? Planta 250:821–838. https://doi.org/10.1007/s00425-018-03084-0

Larcher C, Camerer S (2015) Street Food. Temes de Dissen 31:70–83. Available https://core.ac.uk/download/pdf/39016176.pdf. Accessed 09 Apr 2020

Liu F, Jiao C, Bi C, Xu Q, Chen P, Heuberger AL, Krishnan HB (2017) Metabolomics approach to understand mechanisms of β-N-oxalyl-l-α, β-diaminopropionic acid (β-ODAP) biosynthesis in grass pea (*Lathyrus sativus* L.). J Agric Food Chem 65(47):10206–10213. https://doi.org/10.1021/acs.jafc.7b04037

López Aydillo NR, Toledano Jiménez Castellanos A (1968) Contribucion a la etiologia y patogenia del latirismo experimental en ratones blancos mediante laingestion exclusiva de harna de almortas (Lathyrus sativus) y dietas en blanco, mixta y ajustadas total y parcialmente. Discusion de los factores carencial y neurotoxico desde el punto de vista clinico e histopatologico. Trab Inst Cajal Invest Biol 60:157–190

Manning JA (2009) Constantly containing. Dissertation, West Virginia University

Maranzano B (2014) Lo sviluppo del fenomeno "street food": il cibo di strada a Palermo ieri e oggi. Dissertation, University of Pisa, Italy

McHugh MR (2015) Modern Palermitan Markets and Street Food in the Ancient Roman World. Conference paper, the Oxford Symposium on Food and Cookery, St. Catherine's College, Oxford University

Morton PE (2014) Tortillas: a cultural history. University of New Mexico Press, Albuquerque

Nzwalo H, Cliff J (2011) Konzo: from poverty, cassava, and cyanogen intake to toxico-nutritional neurological disease. PloS Negl Trop Dis 5(6):e1051. https://doi.org/10.1371/journal.pntd.0001051

Patel K, Guenther D, Wiebe K, Seburn RA (2014) Promoting food security and livelihoods for urban poor through the informal sector: a case study of street food vendors in Madurai, Tamil Nadu, India. Food Sec 6(6):861–878. https://doi.org/10.1007/s12571-014-0391-z

Pilcher JM (2017) Planet taco: a global history of Mexican food. Oxford University Press, Oxford

Pisano A (2011) La farinata diventa fainé. Un esempio di indigenizzazione. Intrecci. Quaderni di antropologia cultural I, 1:35–58. Associazione Culturale Demo Etno Antropologica, Sassari

Sheen B (2010) Foods of Egypt. Greenhaven Publishing LLC, New York

Simopoulus AP, Bhat RV (2000) Street Foods. Karger AG, Basel

Spencer PS, Palmer VS (2003) Lathyrism: aqueous leaching reduces grass-pea neurotoxicity. Lancet 362(9398):1775–1776

Steven QA (2018) Fast food, street food: western fast food's influence on fast service food in China. Dissertation, Duke University, Durham

Teklehaimanot R, Wuhib E, Kassina A, Kidane Y, Alemu T, Spencer PS (1993) Patterns of *Lathyrus sativus* (grass pea) consumption and ODAP content of food samples in the lathyrism endemic regions of North West Ethiopia. Nutr Res 3:1113–1126

Teklehaimanot T, Feleke A, Lambein F (2005) Is lathyrism still endemic in northern Ethiopia?— The case of Legambo Woreda (district) in the South Wollo Zone, Amhara National Regional State. Ethiop J Health Develop 19:230–236

Webb RE, Hyatt SA (1988) Haitian street foods and their nutritional contribution to dietary intake. Ecol Food Nutr 21(3):199–209. https://doi.org/10.1080/03670244.1988.9991033

Wilkins J, Hill S (2009) Food in the ancient world. Blackwell Publishing Ltd., Maiden, Oxford, and Carlton

Wilkinson J (2004) The food processing industry, globalization and developing countries. Electron J Agric Develop Econ 1(2):184–201

Xie GX, Qiu YP, Qiu MF, Gao XF, Liu YM, Jia W (2007) Analysis of dencichine in Panax noto-ginseng by gas chromatography–mass spectrometry with ethyl chloroformate derivatization. J Pharm Biomed Anal 43(3):920–925. https://doi.org/10.1016/j.jpba.2006.09.009

Yan ZY, Spencer PS, Li ZX, Liang YM, Wang YF, Wang CY, Li FM (2006) Lathyrus sativus (grass pea) and its neurotoxin ODAP. Phytochem 67(2):107–121. https://doi.org/10.1016/j.phytochem.2005.10.022

Chapter 5
Traditional Grass Pea-Based Foods. Authenticity and Traceability Concerns

Abstract The production of certain recipes based on both cereals and grain legumes may be supposed to be a good deal when speaking of authenticity and traceability issues. Apparently, grain legumes—and specifically *Lathyrus sativus*—may be included in these discussions when speaking of economically motivated adulteration. Actually, safety and health concerns related to grass pea should be good arguments against food frauds linked with the commerce of *L. sativus*. However, the Italian perspective suggests some counterargument. Several cultivars are well known and considered 'Traditional Agricultural Food Products' in Italy, meaning that the historical tradition should be correlated with the geographical importance. Anyway, two different viewpoints have to be considered when speaking of authenticity and traceability issues in connection with grass pea: first of all, the birth or growth of 'market niches' with a strong interest in grass pea, in countries with high income; and secondly, the possible use of this grain legume, with the intention of adulterating other flours, in some Asian country above all. For these reasons, grass pea can have some important role when speaking of food frauds.

Keywords Authenticity · Economically motivated adulteration · *Lathyrus sativus* · *Piciocia* · *RASFF* · Traditional agricultural food product · Traceability

Abbreviations

EMA	Economically motivated adulteration
FBO	Food business operator
RAPD	Randomly amplified polymorphic DNA
RASFF	Rapid Alert System for Food and Feed
PAT	Traditional Agricultural Food Product

© The Author(s), under exclusive license to Springer
Nature Switzerland AG 2020
M. Barone and R. Tulumello, *Lathyrus sativus and Nutrition*,
Chemistry of Foods, https://doi.org/10.1007/978-3-030-59091-8_5

5.1 Grass Pea-Based Foods, Authenticity, and Traceability. Two Faces of the Same Medal

The production of certain recipes based on both cereals and grain legumes may be supposed to be a good deal when speaking of authenticity and traceability issues. In other terms, might grain legumes—and specifically *Lathyrus sativus*—be used for economically motivated adulteration?

Apparently, the answer should be negative based on safety and health concerns related to this grain legume (grass pea), as discussed in Chap. 4. On the other side, grass pea is present in the human history since the Neolithic Era at least. Consequently, it could be considered a cultural heritage of ancient times (Alfiero et al. 2017; Bell and Loukaitou-Sideris 2014; Booth and Coveney 2007; Chong and Eun 1992; Cortese et al. 2016; de Suremain 2016; Delgado et al. 2017; Heinzelmann 2014; Helou 2006; Kollnig 2020; Larcher and Camerer 2015; Manning 2009; Maranzano 2014; McHugh 2015; Morton 2014; Patel et al. 2014; Pilcher 2017; Sheen 2010; Simopoulus and Bhat 2000; Steven 2018; Webb and Hyatt 1988; Wilkins and Hill 2009; Wilkinson 2004).

The Italian perspective concerning grass pea can be interesting enough. Several grass pea food products are well known and considered 'Traditional Agricultural Food Products' (PAT) in Italy, according to the document 'Decreto Ministeriale 18 luglio 2000', 'Elenco nazionale dei prodotti agroalimentari tradizionali' (Ministero delle politiche agricole alimentari e forestali 2020).

The Italian name of grass pea, *cicerchia*, is reported several times with reference to different Italian regions, meaning that the historical tradition should be correlated with the geographical importance. Also, some project has been announced in past years concerning one specific grass pea-based food, the *Piciocia* of Troina (Sicily) and the possible proposal as PAT, in the ambit of an agreement between the City of Troina and the Interprovincial Order of Chemists of Sicily (Città di Troina 2017). More recently, an agreement concerning *piciocia* has been announced between the City of Troina and Slow Food Sicilia, demonstrating the perceived importance of this grass pea-based product for the local community and related economy (Inveninato 2019). Different historical recipes based on cereals and also grain legumes such as chickpeas are recognised as valuable heritages of past cultures (Abulafia 1978; Barbagli and Barzini 2010; Carrabino 2017; CNC 2014a, b, c, d, e; Dauverd 2006; Guigoni 2004; Pisano 2011).

With reference to the Italian situation (Fig. 3.1), *L. sativus* appears to be considered as PAT or as ingredient of a PAT food in the following regions, although some correction has to be made:

1. Apulia: grain legume, named *Cicerchia, Fasul a gheng, Cicercola, Cece nero, Ingrassamanzo, Dente di vecchia,* or *Pisello quadrato*

2. Abruzzo:

    2.1 Delicatessen, named *'cicerchiata'*. Interestingly, this food refers to grass pea (*cicerchie*) with relation to its visual appearance, while grass pea is not used at all

    2.2 Pasta preparation: *'Sagne a pezze e cicerchie'* (Anonymous 2020). This cereal-based food refers to grass pea (*cicerchie*) but *L. sativus* is not used at all

3. Lazio: vegetable product, *'cicerchia'* (grass pea 'as it is')
4. Marche:

    4.1. Vegetable product, *'cicerchia'* (grass pea 'as it is')

    4.2. Delicatessen, named *'cicerchiata'*. Interestingly, this food refers to grass pea (*cicerchie*) with relation to its visual appearance, while grass pea is not used at all

5. Molise:

    5.1. Vegetable product, *'cicerchia'* (grass pea 'as it is')

    5.2. Delicatessen, named *'Cicerchiata'*. This food refers to grass pea (*cicerchie*) with relation to its visual appearance, while *L. sativus* is not used at all

6. Sardinia: vegetable product, *'cicerchia sarda'* (grass pea 'as it is')
7. Umbria: vegetable product, 'cicerchia' (grass pea 'as it is').

As above mentioned, some interest for grass pea and historical *L. sativus*-based recipes can be found in Italy, and probably in other areas worldwide. The question is: Could grass pea be related with authenticity or traceability issues, in connection with food adulteration?

The answer has to be considered by two different viewpoints:

1. The birth or growth of 'market niches' with a strong interest in grass pea and its positive features (also by the historical viewpoint). This phenomenon should be observed in countries with high income.
2. The possible use of grass pea, generally as flour, with the intention of adulterating other flours. This situation is already observed, in some Asian country above all.

5.2 Authentic and Traceable Grass Pea and Related Recipes. Market Niches

In general, the commerce of grass pea and related recipes appears limited to a few markets and food business operators (FBO) only. Naturally, the 'bad' reputation and recommendations for preparation and use remain the main reason for this situation. On the other hand, *L. sativus* appears also to have a certain interest in the market of organic foods, especially if definitions such as the Italian PAT (Sect. 5.1) can be achieved and used.

In Italy, there is a good market niche concerning only organic *cicerchia*, while grass pea-based recipes remain only in the ambit of local traditions, and consequently prepared and served by local catering services or restaurants, or during festival periods (Lambein et al. 2019). As a consequence, grass pea is not a general-use grain legume, in countries with high or medium income at least. Naturally, the situation is also dependent on the long absence of famine periods in Europe: in fact, grass pea was an important crop during the Spanish Civil War (Lambein et al. 2019).

Another possible market is the animal nutrition, with specific relation to organically produced animals, as reported recently (Baldinger et al. 2016). The common point is the use in the production of organic foods, and this fact highlights the importance of this legume not only with reference to its nutritional profile, but also with relation to grass pea as peculiar (uncommon) organic feed.

It has to be noted that the European 'Rapid Alert System for Food and Feed' (RASFF) Portal has not important notifications concerning grass pea when speaking of foods imported in the European Union. The only recorded episode—reference No. 2006.0770, date: 06th November 2006—concerned the withdrawal from the European market of neurotoxic grass pea (*L. sativus*) seeds from Italy to Denmark (RASFF 2020). Substantially, the reason of withdrawal was only linked to neurotoxic properties of grass pea, while no fraud issues were considered. After this episode, grass pea has not considered and recorded in the RASFF Portal. On these bases, it might be concluded that there are not serious reasons for suspecting economically motivated adulteration (EMA) with relation to declared grass pea, even if this grain legume has a certain importance in the market of organic foods.

In addition, grass pea has been recently proposed as a healthy nutraceutical food (Lambein et al. 2019). However, the situation is not apparently modified in recent times, from the viewpoint of markets. It has to be considered that the simple looking of grass pea grains is apparently sufficient when speaking of a simple investigation. On the contrary, grain legumes might be turned in flours and consequently added to grass pea flours (or named as 'grass pea' flours). However, this possibility has not been observed at present, with the exception of one reported case in Ethiopia at least (Woldeamanuel et al. 2012), while the opposite possibility is really a problem.

5.3   Grass Pea as Undeclared Grain Legume in Flours. Real Food Frauds

It has been frequently reported that legume flours may be adulterated with grass pea flour. Generally, adulterated flours are chickpea or Bengal gram (*Cicer arietinum*) and pigeon pea or redgram (*Cajanus cajan*) (Gahukar 2014; Haimanot et al. 2007; Urga et al. 1995, 2005). Grass pea flours are reported to be used as fraudulent additives for legume flours such as chick peas, dry peas, and Bengal gram. This adulteration is extremely worrying because of the high rate of substitution on the one hand, and obvious health and safety problems related to the unknown (and sometimes banned) *L. sativus* presence (heat treatments can be used but without detailed information) (Dixit et al. 2016; Haimanot et al. 2007; Teklehaimanot et al. 1993; Woldeamanuel et al. 2012; Yadav 1996).

Generally, reported cases of adulteration of mixed flours with grass pea are observed in African or Asian countries, including India, Ethiopia, and Nepal, with the aim of preparing flours of subsequent use in meals and snacks (Arora et al. 1996; Gahukar 2014; Haimanot et al. 2007; Teklehaimanot et al. 1993). Unfortunately, the problem is that high-quality flours can be mixed with grass pea flours without the possibility of detecting visually important differences. For these reasons, many analytical methods have been created with concern to the detection of the neurotoxin responsible for neurolathyrism (the most important safety concern). Randomly amplified polymorphic DNA (RAPD) fingerprint analysis may be used for large screening studies concerning *L. sativus* genetic features. Anyway, there are traceability and authenticity concerns in this field, with relation to grass pea as adulterant, instead of the opposite situation (Lioi et al. 2011; Tsegaye et al. 2007). Clearly, more research is needed in the next years because of current and next globalisation challenges.

References

Abulafia D (1978) Pisan commercial colonies and consulates in twelfth-century Sicily. Eng Historical Rev 93(366):68–81

Alfiero S, Giudice AL, Bonadonna A (2017) Street food and innovation: the food truck phenomenon. Brit Food J 119(11):2462–2476. https://doi.org/10.1108/BFJ-03-2017-0179

Anonymous (2020) Sagne a pezze e cicerchie. Top Food Italy. Available https://www.topfooditaly.net/prodotto/sagne-pezze-cicerchie/?v=2a47ad90f2ae. Accessed 14 July 2020

Arora RK, Mathur PN, Riley KW, Adham Y (1996) Lathyrus genetic resources in Asia. In: Proceedings of a regional workshop, 27–29 Dec 1995, Indira Gandhi Agricultural University, Raipur, India, pp 27–29

Baldinger L, Hagmüller W, Minihuber U, Schipflinger M, Zollitsch W (2016) Organic grass pea (*Lathyrus sativus* L.) seeds as a protein source for weaned piglets: effects of seed treatment and different inclusion rates on animal performance. Renew Agric Food Sys 31(3):269–272. https://doi.org/10.1017/S1742170515000186

Barbagli A, Barzini S (2010) Pane, pizze e focacce. Giunti Editore, Florence

Bell JS, Loukaitou-Sideris A (2014) Sidewalk informality: an examination of street vending regulation in China. Int Plan Stud 19(3–4):221–243. https://doi.org/10.1080/13563475.2014.880333

Booth SL, Coveney J (2007) Survival on the streets: prosocial and moral behaviors among food insecure homeless youth in Adelaide, South Australia. J Hunger Environ Nutr 2(1):41–53. https://doi.org/10.1080/19320240802080874

Carrabino D (2017) 14 Oratories of the Compagnie of Palermo: sacred spaces of rivalry. In: Bullen Presciutti D (ed) Space, place, and motion: locating confraternities in the late medieval and early modern city. Koninklijke Brill Nv (Brill), Leiden, pp 344–371. https://doi.org/10.1163/9789004339521_016

Chong HK, Eun VLN (1992) 4 backlanes as contested regions: construction and control of physical. In: Huat CB, Edwards N (eds) Public space: design, use and management. Singapore University Press, Singapore

Città di Troina (2017) Sottoscritto protocollo d'intesa con il Consiglio Nazionale dei Chimici per la salvaguardia e la garanzia della tradizione gastronomica di Troina. Available https://www.com une.troina.en.it/DETTAGLIO_NEWS.ASP?ID=470. Accessed 14 July 2020

CNC (2014a) Accordo tra Comune di Palermo e Consiglio Nazionale dei Chimici (CNC) per la partecipazione alla realizzazione di un sistema di salvaguardia e garanzia della tradizione gastronomica palermitana', Prot. 646/14/cnc/fta. Consiglio Nazionale dei Chimici (CNC), Rome. Available https://www.chimicifisici.it/wp-content/uploads/2018/10/20131210_accordo_firmato_dal_Presidente_del_CNC.pdf. Accessed 07 Apr 2020

CNC (2014b) DOMANDA DI REGISTRAZIONE—Art. 8—Regolamento (UE) n. 1151/2012 del Parlamento Europeo e del Consiglio del 21 novembre 2012 sui regimi di qualità dei prodotti agricoli e alimentari—"ARANCINA". Annex to the document 'Accordo tra Comune di Palermo e Consiglio Nazionale dei Chimici (CNC) per la partecipazione alla realizzazione di un sistema di salvaguardia e garanzia della tradizione gastronomica palermitana', Prot. 646/14/cnc/fta. Consiglio Nazionale dei Chimici (CNC), Rome. Available https://www.chimicifisici.it/wp-con tent/uploads/2018/10/ARANCINA_CNC__STG_2014.pdf. Accessed 08 Apr 2020

CNC (2014c) DOMANDA DI REGISTRAZIONE—Art. 8—Regolamento (UE) n. 1151/2012 del Parlamento Europeo e del Consiglio del 21 novembre 2012 sui regimi di qualità dei prodotti agri-coli e alimentari—"SFINCIONELLO". Annex to the document 'Accordo tra Comune di Palermo e Consiglio Nazionale dei Chimici (CNC) per la partecipazione alla realizzazione di un sistema di salvaguardia e garanzia della tradizione gastronomica palermitana', Prot. 646/14/cnc/fta. Consiglio Nazionale dei Chimici (CNC), Rome. Available https://www.chimicifisici.it/wp-con tent/uploads/2018/10/SFINCIONELLO_CNC_STG_2014.pdf. Accessed 07 Apr 2020

CNC (2014d) DOMANDA DI REGISTRAZIONE—Art. 8—Regolamento (UE) n. 1151/2012 del Parlamento Europeo e del Consiglio del 21 novembre 2012 sui regimi di qualità dei prodotti agri-coli e alimentari—"PANE CA MEUSA". Annex to the document 'Accordo tra Comune di Palermo e Consiglio Nazionale dei Chimici (CNC) per la partecipazione alla realizzazione di un sistema di salvaguardia e garanzia della tradizione gastronomica palermitana', Prot. 646/14/cnc/fta. Consiglio Nazionale dei Chimici (CNC), Rome. Available https://www.chimicifisici.it/wp-con tent/uploads/2018/10/PANE_CA_MEUSA_rev.3dic2014.pdf. Accessed 08 Apr 2020

CNC (2014e) DOMANDA DI REGISTRAZIONE DI UNA STG—Art. 8—Regolamento (UE) n. 1151/2012 del Parlamento Europeo e del Consiglio del 21 novembre 2012 sui regimi di qualità dei prodotti agricoli e alimentary—"PANE E PANELLE". Annex to the document 'Accordo tra Comune di Palermo e Consiglio Nazionale dei Chimici (CNC) per la partecipazione alla realiz-zazione di un sistema di salvaguardia e garanzia della tradizione gastronomica palermitana', Prot. 646/14/cnc/fta. Consiglio Nazionale dei Chimici (CNC), Rome.Available https://www.chimicifi sici.it/wp-content/uploads/2018/10/PANE_E_PANELLE__CNC_STG_2014.pdf. Accessed 07 Apr 2020

Cortese RDM, Veiros MB, Feldman C, Cavalli SB (2016) Food safety and hygiene practices of vendors during the chain of street food production in Florianopolis, Brazil: a cross-sectional study. Food Control 62:178–186. https://doi.org/10.1016/j.foodcont.2015.10.027

Dauverd C (2006) Genoese and catalans: trade diaspora in early modern Sicily. Mediterr Stud 15:42–61

de Suremain CÉ (2016) The never-ending reinvention of 'traditional food'. In: Sébastia B (ed) Eating traditional food: politics, identity and practices. Routledge, Abingdon

Delgado AM, Almeida MDV, Parisi S (2017) Chemistry of the Mediterranean Diet. Springer International Publishing, Cham. https://doi.org/10.1007/978-3-319-29370-7

Dixit GP, Parihar AK, Bohra A, Singh NP (2016) Achievements and prospects of grass pea (*Lathyrus sativus* L.) improvement for sustainable food production. Crop J 4(5):407–416. https://doi.org/10.1016/j.cj.2016.06.008

Gahukar RT (2014) Food adulteration and contamination in India: occurrence, implication and safety measures. Int J Basic Appl Sci 3(1):47–54. https://doi.org/10.14419/ijbas.v3i1.1727

Guigoni A (2004) La cucina di strada Con una breve etnografia dello street food genovese. Mneme-Revista de Humanidades 4(9). fev./mar.de2004:32–43

Haimanot RT, Kidane Y, Thippeswamy R, Martin A, Gowda LR (2007) A reverse phase high performance liquid chromatography method for analyzing of neurotoxin β-N-oxalyl-l-α, β-diaminopropanoic acid in legume seeds. Food Chem 101(3):1290–1295. https://doi.org/10.1016/j.foodchem.2005.12.024

Heinzelmann U (2014) Beyond bratwurst: a history of food in Germany. Reaktion Books Ltd., London

Helou A (2006) Mediterranean street food: stories, soups, snacks, sandwiches, barbecues, sweets, and more from Europe, North Africa, and the Middle East. Harper Collins Publishers, New York

Inveninato R (2019) Troina. Nasce la comunità del cibo Slow Food. Enna Magazine. Available https://www.ennamagazine.it/Comuni/TabId/102/ArtMID/587/ArticleID/3513/Troina-Nasce-la-comunit224-del-cibo-Slow-Food.aspx#.Xw2T08dR21t. Accessed 14 July 2020

Kollnig S (2020) The 'good people' of Cochabamba city: ethnicity and race in Bolivian middle-class food culture. Lat Am Caribb Ethn Stud 15(1):23–43. https://doi.org/10.1080/17442222.2020.169 1795

Lambein F, Travella S, Kuo YH, Van Montagu M, Heijde M (2019) Grass pea (Lathyrus sativus L.): orphan crop, nutraceutical or just plain food? Planta 250:821–838. https://doi.org/10.1007/s00425-018-03084-0

Larcher C, Camerer S (2015) Street Food. Temes de Dissen 31:70–83. Available https://core.ac.uk/download/pdf/39016176.pdf. Accessed 09 Apr 2020

Lioi L, Sparvoli F, Sonnante G, Laghetti G, Lupo F, Zaccardelli M (2011) Characterization of Italian grasspea (Lathyrus sativus L.) germplasm using agronomic traits, biochemical and molecular markers. Genet Res Crop Evol 58(3):425–437. https://doi.org/10.1007/s10722-010-9589-x

Manning JA (2009) Constantly containing. Dissertation, West Virginia University

Maranzano B (2014) Lo sviluppo del fenomeno "street food": il cibo di strada a Palermo ieri e oggi. Dissertation, University of Pisa, Italy

McHugh MR (2015) Modern palermitan markets and street food in the ancient roman world. In: Conference paper, the Oxford Symposium on Food and Cookery, St. Catherine's College, Oxford University

Ministero delle politiche agricole alimentari e forestali (2020) Decreto ministeriale 18 luglio 2000, "Elenco nazionale dei prodotti agroalimentari tradizionali", 20th revision, published on the 'Gazzetta Ufficiale della Repubblica Italiana', Serie Generale n.42 of 20–02–2020. Ministero delle politiche agricole alimentari e forestali, Rome. Available https://www.politicheagricole.it/flex/cm/pages/ServeBLOB.php/L/IT/IDPagina/2769. Accessed 14th July 2020

Morton PE (2014) Tortillas: a cultural history. University of New Mexico Press, Albuquerque

Patel K, Guenther D, Wiebe K, Seburn RA (2014) Promoting food security and livelihoods for urban poor through the informal sector: a case study of street food vendors in Madurai, Tamil Nadu, India. Food Sec 6(6):861–878. https://doi.org/10.1007/s12571-014-0391-z

Pilcher JM (2017) Planet taco: a global history of Mexican food. Oxford University Press, Oxford

Pisano A (2011) La farinata diventa fainé. Un esempio di indigenizzazione. Intrecci. Quaderni di antropologia cultural I, 1:35–58. Associazione Culturale Demo Etno Antropologica, Sassari

RASFF (2020) RASFF Portal, Notification details-2006.0770. European Commission, Brussels. Available https://webgate.ec.europa.eu/rasff-window/portal/?event=notificationDetail&NOTIF_REFERENCE=2006.0770. Accessed 14th July 2020

Sheen B (2010) Foods of Egypt. Greenhaven Publishing LLC, New York

Simopoulus AP, Bhat RV (2000) Street foods. Karger AG, Basel

Steven QA (2018) Fast food, street food: western fast food's influence on fast service food in China. Dissertation, Duke University, Durham

Teklehaimanot R, Wuhib E, Kassina A, Kidane Y, Alemu T, Spencer PS (1993) Patterns of Lathyrus sativus (grass pea) consumption and ODAP content of food samples in the lathyrism endemic regions of North West Ethiopia. Nutr Res 3:1113–1126

Tsegaye M, Demissew S, Alexandra J (2007) Assessment of diversity, morphological variation and description of Grasspea (*Lathyrus sativus*) and other related species. Dissertation, Addis Ababa University, Addis Ababa

Urga K, Fite A, Kebede B (1995) Nutritional and anti-nutritional factors of grass pea germ plasm. Bull Chem Soc Ethiop 9:9–16

Urga K, Fufa H, Biratu E, Husain A (2005) Evaluation of *Lathyrus sativus* cultivated in Ethiopia for proximate composition, minerals, β-ODAP and anti-nutritional components. Afric J Food Agric Nutr Dev 5(1):1–15. Available https://www.ajol.info/index.php/ajfand/article/view/135979. Accessed 08th July 2020

Webb RE, Hyatt SA (1988) Haitian street foods and their nutritional contribution to dietary intake. Ecol Food Nutr 21(3):199–209. https://doi.org/10.1080/03670244.1988.9991033

Wilkins J, Hill S (2009) Food in the ancient world. Blackwell Publishing Ltd., Maiden, Oxford, and Carlton

Wilkinson J (2004) The food processing industry, globalization and developing countries. Electron J Agric Dev Econ 1(2):184–201

Woldeamanuel YW, Hassan A, Zenebe G (2012) Neurolathyrism: two Ethiopian case reports and review of the literature. J Neurol 259(7):1263–1268. https://doi.org/10.1007/s00415-011-6306-4

Yadav CR (1996) Genetic evaluation and varietal improvement of grasspea in Nepal. In: Proceedings of a regional workshop, 27–29 Dec 1995, Indira Gandhi Agricultural University, Raipur, India, p 21

Printed in the United States
By Bookmasters